Continental Philosophy of Social Science

Hermeneutics, Genealogy and Critical Theory from Ancient Greece to the Twenty-First Century

This is the first English-language book dedicated to the Continental tradition in the philosophy of social science. It seeks to demonstrate the unique nature of the continental approach to the philosophy of social science and contrast this with the Anglo-American rubric. The subject matter spans the traditions of hermeneutics, genealogy and critical theory, examining the key texts and theories of thinkers like Gadamer, Ricoeur, Derrida, Nietzsche, Foucault, the Early Frankfurt School and Habermas. The approach is highly original in that it shows the historical depth of these mainstays of twentieth-century thought by tracing their ideas back to origins in Ancient Greece and Rome, medieval Christian thought, the Enlightenment and Romanticism. Throughout, Yvonne Sherratt argues for the importance of historical understanding in order to appreciate the distinct, humanist character of Continental philosophy of social science. This book will form the essential bedrock of any course in the philosophy of the social or human sciences. It is also an essential counterpoint to extant texts in the field and has widespread inter-disciplinary appeal.

Yvonne Sherratt is British Academy Researcher in the Department of Philosophy at the University of Oxford. She has taught at the Universities of Cambridge, Edinburgh and Wales, and she is the author of *Adorno's Positive Dialectic*.

Continental Philosophy of Social Science

Hermeneutics, Genealogy and Critical Theory from Ancient Greece to the Twenty-First Century

YVONNE SHERRATT

University of Oxford
and
University of Wales

CAMBRIDGE UNIVERSITY PRESS
Cambridge, New York, Melbourne, Madrid, Cape Town, Singapore, São Paulo

Cambridge University Press
40 West 20th Street, New York, NY 10011-4211, USA

www.cambridge.org
Information on this title: www.cambridge.org/9780521854696

© Yvonne Sherratt 2006

First published 2006

Printed in the United States of America

A catalog record for this publication is available from the British Library.

Library of Congress Cataloging in Publication Data

Continental philosophy of social science : hermeneutics, genealogy and critical theory
from Ancient Greece to the twenty-first century / Yvonne Sherratt.
p. cm.
Includes bibliographical references (p.   ) and index.
ISBN 0-521-85469-5 (hardback) – ISBN 0-521-67098-5 (pbk.)
1. Hermeneutics – History.   2. Genealogy (Philosophy) – History.
3. Critical theory – History.   4. Social sciences – Philosophy – History.   I. Title.
BD241.S497   2006
190 – dc22   2005006334

ISBN-13   978-0-521-85469-6 hardback
ISBN-10   0-521-85469-5 hardback

ISBN-13   978-0-521-67098-2 paperback
ISBN-10   0-521-67098-5 paperback

For Clara.

Contents

Preface

I would like to thank Raymond Geuss for initial inspiration for many of the topics in this book. I owe a debt to Mark Peacock Lambert Zuidervaart and Nigel Pleasants for their insightful comments on various parts of this book. My great thanks go to Robert Mayhew for reading the entire script. I would also like to acknowledge a debt of gratitude to the National Library of Wales for the use of its resources, to Beatrice Rehl and Laura Lawrie at Cambridge University Press and to the late Terence Moore for helping initiate this project, although he was sadly unable to see it to completion. This book was written during an extended period of maternity leave and is dedicated to my new baby daughter Clara.

Continental Philosophy of Social Science

Hermeneutics, Genealogy and Critical Theory from Ancient Greece to the Twenty-First Century

Introduction

Astonishingly there exists no single English language book dedicated to the continental tradition in the philosophy of social science. Juxtapose this surprising position with the abundance of continental works on all other aspects of European thought from the vast array of tracts on political philosophy, aesthetics and the history of European ideas to the plethora of studies of individual thinkers and schools of thought. Contrast, too, this lack of a treatment of continental philosophy of social science with the abundance of expositions of the same subject matter in the Anglo-American tradition.

In fact, the entire mainstream corpus of philosophy of social science is dominated by Anglo-American literature. Consequently these transatlantic schools set the agenda for pretty much all philosophy of social science with their own particular approach to the subject matter – a distinct canon of thinkers, set of questions and debates. The continental domain is simply marginalized by this rubric: either through sheer exclusion or, indeed, by a reductive mode of inclusion.

On the one hand, in so far as continental issues veer from the Anglo-American agenda, the continental tradition is simply ostracised. On the other hand, in so far as continental questions address this 'mainstream' rubric the tradition is included. However this means that in so far as the continental philosophy of social science exists at all, it is merely treated as a strand of influence in the supposedly much larger world of Anglo-American approaches. That is to say, European ideas are subsumed as influences upon and contributions to the greater project of Anglo-American philosophy of social science.

Continental philosophy of social science however is more than just a part of a larger Anglo-American corpus. It is a distinguished and autonomous strand of thought in its own right. Continental schools have their own canon of thinkers, pose their own questions, set their own agendas and have a rich, deep history stemming back to Ancient Greece, Rome and early Christendom. In fact, it is this connection to its Ancient past, let us say, its *humanism*, that defines the continental tradition.

This link with humanism is what sets the continental schools of thought apart from Anglo-American philosophy of social science. Moreover, it is the maintenance of a *living* connection with the humanist past that creates a problem for the Anglo-Americans, for this latter rubric has discarded many humanist concerns in favour of the issues of science. In so far as it pursues any philosophical rebellion against science at all, the Anglo-American agenda does so from a 'post-scientific' perspective in which earlier humanism is still completely forgotten.

My contention is that continental philosophy of social science is best understood as emergent from humanism. The aim of this book is twofold; first, to offer an account of philosophy of social science dedicated to the continental tradition. Second, we wish to depict the autonomous nature of the continental field of enquiry by exploring its special humanist history, and highlight the distinct approach to philosophy of social science that has arisen as a consequence of this.

Before going on to explore continental philosophy of social science, let us first undertake to clarify a few key terms. A preliminary and fairly self-evident point to note is that the term 'continental' is defined in contrast to Anglo-American philosophy. The distinction between these traditions is very marked. Whilst the latter is often an analytical, concept-based style of analysis with little regard for historical factors, continental philosophy is usually a text-centred, historically sensitive tradition. These approaches also embrace different canons and address diverse philosophical questions. The specificities of the distinctions between the continental and Anglo-American approaches when it comes to the particular area of philosophy of *social science* will be discussed in more detail throughout this study.

A further minor point is that we use the notion 'continental' in contrast with the word 'European'. Our reason for this is really that continental tends to be more frequently used to refer to concerns that stem from *Continental* Europe as opposed to Britain. Strictly speaking 'European' embraces British philosophy, too, although the term is most often used to capture the distinct continental traditions and exclude British ones which are usually analytical.

Our final term seems an obvious one, namely, what we mean by philosophy of social science. However, the definition of this notion becomes central in that it is perhaps the main reason why no specialist study in continental philosophy of social science exists. Philosophy of social science as we have said is defined in its mainstream rubric to marginalize many of the main concerns of the continental tradition. It is conceptualised to centre around the kind of questions dear to Anglo-American philosophers. Mainly it is concerned with the relationship between the natural and the social sciences and the sorts of complexities that arise from studying society as a possible object of scientific enquiry. Furthermore, it is a comparatively recent discipline with little historic focus. If we take this Anglo-American view of the discipline, our Continental study is barely a philosophy of social science at all. However, if we embrace a broader notion of social science as pertaining to all the traditions engaged in the study of human society, both recent and much older, then the philosophy of social science becomes the tradition of philosophy that addresses the problems and techniques involved in studying human society. Within this broader, more *humanist* conception, the continental tradition occupies a pivotal place.[1]

Let us now contextualise our own specifically continental study within the other literature on philosophy of social science. There exists a wealth of primary and secondary literature delimiting a well-researched, often sophisticatedly analysed and hotly debated domain. The literature is strongly skewed towards the Anglo-American traditions – although this is not to underestimate the diversity of the Anglo-American canon. Within this diversity certain patterns of analysis emerge. The majority of treatments are 'modern' and almost all accounts of philosophy of social science fall within two main strands. They encapsulate either what we can describe as a 'social sciences led approach' or what we can demarcate as a 'philosophically animated' one. Less than a handful of the overall philosophy of social science literature presents any historical depth and none explores the continental tradition as its main focus.

First, let us consider in more detail the first group of studies, namely the Anglo-American social sciences led approach to philosophy of social science. We perceive studies derived from disciplines concerned with empirical issues and often conducted by those with empirical training. Indeed, these tracts often emanate from those educated within the social sciences – usually sociologists – although some studies emerge from those with a natural sciences backgrounds. These works are ubiquitously

---

[1] We define what we mean by humanism later in this introduction.

modern and preoccupied with questions of the relationship between the social and natural sciences.

These social sciences–derived approaches can further be categorised as broadly empiricist or post-empiricist according to when they were written. To those studies conducted prior to the 1950s and 1960s we can attribute a firmly empiricist mentality. As we all know, in the seventeenth and eighteenth centuries, the empiricist and rationalist traditions were at their peak. This was the era of the dawning of modern science and philosophy of social science never looks beyond this era for its basic philosophical assumptions. Often, however, these scholars look to the much more historically recent nineteenth-century empiricist traditions,[2] positivism being an obvious source of inspiration,[3] and many indeed are concerned only with issues rising subsequent to the first half of the twentieth century.

These empiricist philosophers of social science are writing during an era in which the authority of science is fairly absolute. Given the youth of the *social* sciences, they are concerned with establishing the legitimacy of their discipline and of formulating foundational methodologies. Thus these empiricist philosophers were seeking to mesh the successful methodologies from natural science onto social science. Their works seek to establish a clear rubric for their field and demonstrate a clear methodology. Philosophy of social science, therefore, embraces many of the concerns of the natural sciences, for instance, issues of experimentation, causality, prediction, explanation, how particularities can be subsumed under general laws. This Anglo-American empiricist philosophy of social science seeks to show how objectivity can be attained over the subjectivity of observers and how to address the fact/value distinction in relation to society as the object of enquiry. Moreover it analyses society through concepts of individuals and the whole, structure and function, action and development and it debates the importance of the economic or psychological, the macro or the micro as the main determining factor. These concerns, whilst representative of the whole tradition of early and mid-twentieth century philosophy of social science have also been carried forwards as forming much of the rubric of the contemporary discipline.[4]

---

[2] See for example, Turner 1986; and Runciman's 1969 and 1972 excellent accounts.

[3] See Bryant 1985, Halfpenny 1982.

[4] See Azevedo 1997, Benton and Craib 2001, Craib 1992 and 1997, Doyal 1986, Galvotti 2003, Glassner and Moreno 1989, Hollis 1994, 1996; Hookway and Pettit 1978, Rosenberg 1988, Shoemaker, Tankard and Lasorsa 2004, Stretton 1969 and Tudor 1982, for earlier and later twentieth century discussions of this rubric of philosophy of social science.

The second group of social sciences led approaches to philosophy of social science can be described as 'post-empiricist'.[5] These studies written during the latter part of the twentieth century occur against a different 'ideological' background. Science no longer reigns unchallenged. Various constructivist, relativist and post-modernist developments challenge the philosophical authority of the natural sciences. The social sciences, still emulating their giant cousin, need to take on board these challenges, too. They respond in various ways. First certain studies aim for impartiality in the empiricist/post-empiricist debate and simply continue to discuss the complexities of the relationship between the social and natural sciences.[6] Whilst these studies include discussions of the new complexities of practising any science, they are still primarily concerned with the relationship between natural and social science so the empiricist rubric is fairly unchanged.

A further group of studies from the post-empiricist era jettison impartiality and firmly attempt to restore empiricist values. Within this group of social science–led studies, there are those who bravely try to defend the positivism of the earlier era.[7] More common are those who attempt to redefine science itself and abandon positivism for a new form of science, which they elaborate as critical realism.[8] Others attempt to bolster scientific approaches to the study of society by drawing upon arguments from biology, most notably Darwinism.[9] Many other studies attempt to accommodate the natural and social sciences by revising either the former or the latter.[10]

A final high-profile group of social science–led philosophers of social science are post-empiricist in mentality as well as era. They seek to reject the natural sciences as the model for the social sciences.[11] These studies often look to a strand of continental thought as providing the ammunition to break new post-empiricist approaches. Some turn to the influence of thinkers like Foucault and incorporate social constructivist arguments.[12] Others borrow from post-modernism.[13] A further group

[5] See Agassi et al. 1995; Bohman 1991; Cohen and Wartofsky eds. 1983; Keat and Urry 1982; Kukla 2000; Phillips 1987; Potter 2000; Thomas 1979.
[6] Some of these studies form a continuum with earlier works mentioned. Galvotti 2003, Hollis 1994, 1996; Phillips 1987, Rudner 1966, Wisdom 1993.
[7] See Agassi, ed 1995; Cohen 1993.
[8] Bhaskar 1978; Keat and Urry 1982; Outhwaite 1987.
[9] See Dupre 2003; Rosenberg 2000.
[10] Azevedo 1997.
[11] Hindess 1977 is a good example.
[12] Kukla 2000.
[13] Hindess 1977; Potter 2000.

turn to the omnipresent influence of Marxism,[14] phenomenology[15] or indeed hermeneutics.[16] Many look to a more Anglo-American canon of thinkers who incorporate European style concerns with issues of meaning, interpretation or social construction[17] and examine authors like Winch,[18] Kuhn,[19] Feyerband[20] or Lakatos.[21]

Within Anglo-American philosophy of social science there is a second overall approach to the subject matter, namely, what we have previously described as a 'philosophically animated' one. This philosophical approach stems, as one might expect, from those practicing from within the discipline of philosophy itself. This approach is, therefore, more conceptual and concerned with broad questions derived from those with 'rational', theoretical rather than empirical concerns. The style of these philosophical accounts is analytic so analysis proceeds through concepts and analytical categories. This is in contrast to any historical or textual based mode of enquiry that we might find in continental approaches. Moreover, these analytic accounts debate from within their own canon and rubric, which embraces a distinct set of concerns. We can perhaps usefully divide this literature into four separate categories.

First there are general philosophical studies in which debates from empiricism and rationalism abound. Issues about the nature of the real, debates about fact and idea, the boundary between subjectivity and objectivity and the nature of knowledge preoccupy much of this analysis. Indeed many of the concerns overlap with those from epistemology or metaphysics.[22]

Second, a body of philosophically led philosophy of social science shares the same concerns and language as philosophy of natural science (including the philosophy of biology and psychology).[23] Themes like the nature of explanation, the relevance of natural law, issues of determinism

---

[14] Anderson Hughes and Sharrock 1986; Bottomore 1974; Root 1993; Sayer 1979.
[15] Anderson in Glynn, S., ed. 1986.
[16] Anderson in Glynn, S., ed. 1986.
[17] See Phillips 1987, Bloor 1974, 1976, and Barnes 1974.
[18] Winch 1958.
[19] Kuhn 1962.
[20] Feyerband 1962.
[21] Lakatos and Musgrave 1970.
[22] See Benn and Mortimore 1976; Glynn 1986; Hollis and Lukes, ed. 1982; Martin and Macyntyre 1994; Pratt 1978.
[23] For example of this approach see Bohman 1991; Doyal and Harris 1986; Flew 1985, Hookway and Pettit ed. 1978; Ryan 1970. See also Agassi, Jarvie and Laor, ed. 1995; Cohen, ed. 1983; James's interesting account is perhaps closer to social philosophy but raises some important issues here, see James 1984; or Trigg 2001.

and prediction, questions of objectivity and normativity thus form the mainstay of much of this analysis. Further topics like realism, relativism, absolutism and indeed holism are also pivotal to these works. Uppermost is the question of the appropriateness of natural science for the social sciences and the canon of authors herein bears many similarities to that of the social sciences derived theoretical approach, although the style of analysis here is much more conceptual.[24]

Third, philosophically led philosophy of social science can be founded upon Anglo-American sub-disciplines like philosophy of mind, meta-physics and epistemology.[25] Further fields like political philosophy inform a vast tract of social science works, employing liberal[26] or Marxist arguments and perspectives.[27] Economic philosophy has also made inroads into philosophy of social science with Hayek's work being highly influential.[28] So, too, has philosophy of biology with arguments from Darwinism and ideas from thinkers like Malthus.[29]

Meanwhile, in contrast to these two main strands of philosophy of social science, that is, both the social sciences and philosophically led treatments, there are only a couple of historical studies.[30] One excellent account looks to the origins of a philosophy of social science along with a philosophy of science in Ancient Greece. This impressive scholarly mono-graph, however, focuses only upon the analytic tradition of philosophy of social science, closely following a philosophy of natural sciences rubric, as do any other similar historical studies.[31]

Finally, inside the mainstream Anglo-American rubric some continen-tal ideas have been acknowledged and addressed, often in the form of Marxism, as seen above, or at other times the traditions of hermeneutics, phenomenology or post-modernism might form a chapter in a survey, or a line of influence in an Anglo-American centred debate.[32] Many

[24] For instance, these studies include Popper, Kuhn, Winch, Lakatos, Feyerband and so on. For more sociological perspective on these scientific issues, Barnes 1974 and Bloor 1974, 1976 can usefully be consulted.

[25] See Goldman 1992, Harrison, ed. 1979.

[26] Root 1993.

[27] Cohen 1972; Bottomore 1974; Sayer 1979; Gavroglu, Stachel and Wartofsky ed. 1995.

[28] Hayek 1948, 1952, 1967, 1978; Hollis and Nell 1975.

[29] Rosenberg 2000, Dupré 2003; Malthus 1970.

[30] Compare this lack of a historical treatments in philosophy of social science with the abundance of accounts of the history of the natural sciences, the history of social and political thought and indeed the history of sociology itself.

[31] See especially the excellent, scholarly treatment by Scott 1991; also Rosenberg 1988 and a comprehensive account by Manicas 1987.

[32] Bohman 1991; Glynn, ed. 1986; Anderson Hughes and Sharrock 1986.

Anglo-American philosophers of social science, however, simply ignore the tradition altogether.[33]

What of the continental tradition itself then? Almost any specifically *continental* treatment of philosophy of social science occurs only as an aspect within a more general continental philosophy text, for instance, a tract on continental philosophy or a themed book on Romanticism and so on.[34] Any further treatments in so far as they occur at all do so as part of a single author study.[35] There exist, however, only two works that actually claim to be dedicated to the continental tradition of philosophy of social science itself. One of these is an edited collection of selective, specialised essays rather than an overall study.[36] The second deploys recent continental developments in order to make a critical realist argument and thus subsumes continental ideas within Anglo-American style concerns.[37] There is no study that either examines the continental tradition or explores its humanist past.

Having contextualised a continental account within the mainstream of current philosophy of social sciences literature it now remains for us to elaborate further the scope and nature of our own study. First, we wish to address the lacuna in current philosophy of social science and pursue a specialist treatment of the continental tradition. Our book, we hope, represents a dedicated tract on this neglected area of continental philosophy of social science. Second, we aim to explore the distinctiveness of the European mainland's corpus in contradistinction to its Anglo-American counterpart. We propose to do this by outlining continental philosophy of social science's indebtedness to humanism.

The influence of humanism on continental philosophy of social science impacts on its nature in a variety of ways, all of which make it distinct from the Anglo-American tradition. Let us, therefore, now pursue this overall contention that this corpus is best understood as emanating from and holding many central characteristics of humanism. Having made such a claim, it would be helpful to first pause and briefly explain what we mean by humanism. This is no easy task. Humanism is a fairly broad and descriptive term and one that is not readily condensed into a precise definition. However, I would propose three central points to any definition

---

[33] Flew 1985; Hollis, M. 1994; Trigg 2001.

[34] Critchley 2001 or McNiece 1992.

[35] See for instance Gordon 1995 on Fanon; or any good, comprehensive account of Dilthey, Heidegger or Gadamer.

[36] Babich et al. 1995.

[37] Outhwaite 1987.

of humanism. Moreover I would contrast each of these points with any broadly scientific mentality.

First humanism entails being in touch with the ideas and texts of our *Ancient* ancestors, the Greeks and the Romans. Whether we call ourselves Christians, Romantics or Marxists, if we base our understanding upon a reading of Plato, Aristotle or Cicero and its transmission through the ages, we can rightly lay claim to a humanist stance. This link to our Ancient ancestry contrasts sharply to scientific understanding, which requires no familiarity with the Ancients.

Secondly, humanists hold that knowledge works through *transmission.* Understanding and knowledge are composed by the accumulation of voices handed down from the centuries. This contrasts with science's 'creative destruction' approach where voices from the past are seen as holding false meanings, which need to be destroyed in order to allow new, objective knowledge to flourish. Humanists hold that knowledge comes from the past, which is to say that the past is a source of understanding relevant to the present. This notion includes two distinct but closely related points. On the one hand, progress for humanists would be the accumulation of the knowledge from the past, not the transcendence or destruction of it. Science meanwhile holds the idea that the past contains undeveloped, primitive and indeed often false forms of knowledge. Hence, progress is the replacement of the old by the new. On the other hand, humanism is committed to the idea that it is our prior conceptions that help us build knowledge and understanding. In contrast, science pursues the goal of objectivity, which entails the removal of prior conceptions that are believed to cloud or distort the issue.

Third, humanism I would suggest holds a distinct notion of *meaning* from science. The human world is substantively meaningful for humanists and this *includes* the idea of ethical, aesthetic and even spiritual meanings. Moreover, meaning itself often encompasses the notions of both value and purpose. Society thus for humanists would be an intrinsically purpose-laden, ethically, aesthetically and spiritually valuable entity. This contrasts with a scientific notion of meaning, which is purely technical and pertains only to bare empirical facts. All other forms of human meaning are external, and maybe 'tacked' on as an ethical, aesthetic or indeed subjective addition. Society would be approached as an object of knowledge like any other physical object in the natural world. Humanists also, if you like, hold an ontological assumption about the nature of meaning. The creation of meaning in our world is through human beings. Society, therefore is not only a meaningful entity but one indeed whose meaning is generated through generations of human creativity.[38]

Let us now see just how humanism shapes continental philosophy of social science and generates a distinct tradition for Anglo-American schools of thought. As we have intimated, the continental traditions have their own vivid history stemming back some two thousand years to Ancient Greece and early Christianity. Although Anglo-American philosophy of social science is also justified in looking to humanism for its deeper historical roots, its trajectory differs from its continental counterpart in that it has been subject to a much greater historical *rupture*. The development of the sciences, especially in the seventeenth and eighteenth centuries, and the philosophical arguments that accompanied them, entailed a rejection of many former humanist assumptions and indeed approaches to study of the 'human sciences'. In fact, Anglo-American philosophy of social science looks no further than to the empiricist and rationalist philosophies of the seventeenth and eighteenth centuries for its historical roots. The continental tradition on the other hand, whilst incorporating these developments within its trajectory, does so in such a way that entails no rupture with the earlier, pre-Enlightenment past. It conserves and carries forth an awareness of deeper humanist ideas and scholarship.

This special humanist legacy generates three features to continental philosophy of social science, namely a distinct style of analysis, a particular canon of thinkers and an autonomous set of concerns. First, continental philosophy of social science's distinct history, its living connection with humanism entails a distinct *style of analysis*. It shapes both its style of knowledge acquisition and its assumptions about the nature of the 'object' it wishes to gain knowledge of (society). In contrast to its Anglo-American cousin, continental modes of knowledge acquisition are rooted in the historical, textual and theoretical modes of analysis so dear to earlier humanist scholars. To begin with, history is important and meaningful to continental philosophy of social science. In contrast to Anglo-American styles which revolve around universal concepts which are a-historical, continental thinkers often centre their ideas within historical context. Moreover, whereas Anglo-American philosophies develop arguments around themes generated from concepts, continental approaches pivot around historical traditions and the works of individual thinkers

[38] Note that this position also contrasts with certain Christian views wherein meaning is created through God. However, Christian humanism would hold a more complex fusion of these views.

situated within each historical tradition. Furthermore, continental styles of analysis are more textual than their analytic counterparts. They follow closely a canon of authoritative texts and are sensitive to language and expression. Issues of authorship can play a role and figurative language can be deployed for the purposes of understanding in a way, which mitigates against the more austere analytic tradition. It should be borne in mind, however, that continental styles include philosophical modes of thought deploying concepts. However, this theorising always retains a link to textual or historical issues to a far greater degree than Anglo-American analysis.

Secondly, continental philosophy of social science is shaped by humanism in that it has its own *canon*. Although many of the contemporary thinkers from this, like Gadamer, Habermas or Foucault, have made their way into Anglo-American consciousness and thus into mainstream philosophy of social science, many of their ancestors have not. Greek, Roman and Christian traditions of interpretation are virtually unknown to contemporary philosophers of social science, although they may espouse a deep interest in issues of interpretation.[39] Earlier Enlightenment and Romantic names like Chladenius, Droysen or Humboldt, for example, also rarely figure. Furthermore, critical theory's tentacles into Ancient and Christian ideas have little presence in any Anglo-American philosophical study. Likewise, although Foucault's presence proliferates among many mainstream rubric's of philosophy of social science, students and scholars often show little awareness of Foucault's debt to Nietzsche's classical training and moreover, that Foucault's entire philosophy of social science is built upon Nietzsche's critical engagement with Christianity. Yet all these ideas were crucial to the developing twists and turns of the continental tradition.

Thirdly, this humanist inheritance entails that continental philosophy of social science not only has a distinct style of analysis and a distinct canon of thinkers, but it incorporates its own *intellectual concerns*. This rubric of concerns is vast and varied, but three features perhaps stand out. Ancient thinkers and humanist scholars have always assumed that the object of inquiry, 'society', is meaningful, often linguistic and historical.

The preoccupation with the notion of meaning for continental philosophy of social science is arguably the most important feature derived from humanism and debates rage as to the nature and location of meaning

---

[39] For example, Outhwaite, in a tone of invoking great historical depth, refers to 'hermeneutic theory as being at least 150 years old' 1987: 1.

in human society. Arguments through the ages vary from stating that meaning resides in society's language, its law, its individuals (whether as authors or readers), its humanly constructed gods, a single God, or indeed society's history albeit construed as historical events, ideas, a single moment or the entirety of human history). Further discussions inquire as to exactly how meaning is created and transmitted, whether, for example, it is transmitted through the precision of language, through creative expression or through the mind of the reader or observer. Philosophers enquire if meaning is distorted by human understanding, human relations, power, authority or if indeed meaning resides within such things.

Furthermore, the continental tradition often implicitly takes a notion of society as a historical phenomenon. Therefore, issues about the nature of history are the cornerstone of many debates. Views differ about whether history is changing rapidly, developing slowly, progressing forwards, regressing or merely static. Further questions have been asked if history's change is linear or circular. How is society historical? Is it composed by material or ideal elements; does the way we produce our food impact upon history and shape society or the way we speak and think? Are we internal or external to history, that is, is it something we are part of or are we simply external observers? These and other similar questions form the backbone of many continental questions about the nature of society.

These assumptions of the study of society being about meaning, language and history contrast starkly with Anglo-American approaches, which because of their foundations in philosophy of science, take as their starting point a comparison between the natural and social sciences. Even when not, like positivism or forms of realism, literal attempts to apply science to society, Anglo-American approaches are still formed by the rubric and assumptions of science.

Epistemological questions for the Anglo-American rubric centre upon issues of fact and questions about objectivity of knowledge in contrast to the continental focus upon meaning. For them, issues about proof, prediction and causality replace questions about interpretation and understanding. Debates about whether our knowledge is external to society contrast with the usual continental assumption that understanding is always internal to it.

Ontological topics in the Anglo-American tradition question how like or unlike an object of scientific enquiry society is. Aspects of society are deemed to differ problematically from a concept of pure objectivity – aspects like ethics, the subjectivity of individuals, and so on. Consequently, these are abstractly separated for analysis. Continental views, meanwhile, have never really adopted the ontological assumption that society and its artefacts are like natural objects and should be treated as such. Problems of mapping the non-objective onto the objective, the grand preoccupation of the Anglo-American tradition, therefore, never truly arises in the continental rubric. That is not to say that issues of objectivity do not occur in the continental tradition. They do, but as we have detailed, are questions about the objectivity of linguistic and historical modes of knowledge acquisition and nature of society.

Continental philosophy of social science spans a vast range of thinkers and issues. It encompasses many texts from a range of countries across the continent. Historically, Greece, Rome and Christendom have all bequeathed to the discipline. Its more recent philosophical contributions, however, have come principally from Germany and France, and of these two great nations, it is the former that unarguably forms the bedrock of the discipline and, therefore, is the tradition upon which we most focus.

The project of this book is not to offer a comprehensive survey of all the vast and multifarious contributions of recent times but to show how the major strands of continental philosophy, which form the trunk from which later developments stem, are indebted to humanism. For this reason we have focussed upon the three best examples of transmission of the humanist project through German thought. These are three coherent, unified traditions with great historical depth. We speak of hermeneutics, genealogy and critical theory.[40]

Our three major traditions in continental philosophy of social science uphold all three features of humanism as defined in this introduction. They are all founded upon bodies of knowledge that are either Ancient themselves or engaged with the Ancients. Furthermore, they are corpuses that embody ideas and assumptions accumulated through a canon of voices transmitted over the centuries. Finally, hermeneutics, genealogy and critical theory all maintain the position that society is meaningful and that meaning is, in a variety of ways, created by human beings.

---

[40] This focus entails that some more contemporary strands of continental philosophy have been sidelined. For instance, although we have included phenomenological and post-modern thinkers in so far as they fall within our central historical canons – we discuss Husserl, Heidegger and Derrida under hermeneutics; Foucault under genealogy etc. – we have not included all the recent developments of feminism, post-modernism, current phenomenology or indeed later Habermas. However, by focussing upon the major historical traditions we have provided analysis of the backbone upon which many of these later developments rest, and also detailed their deeper intellectual context.

This book is divided into three parts, part one being concerned with hermeneutics, part two addressing genealogy and the final section, part three, assessing critical theory. Each section explores the history of its particular tradition beginning with Ancient and Christian origins. The discussion then proceeds to continental philosophy of social science and examines contemporary developments within this tradition. Instances of the application of each tradition in social science are included.

Part I begins with an introduction contextualising the main tenets of hermeneutics with respect to other Anglo-American approaches to philosophy of social science. Chapter 1 explores Ancient hermeneutics beginning with a discussion of the earliest uses of the term in Greek contexts and examines interpretation's early connection with myth, grammar and rhetoric. We then look at Roman innovations and the effect upon hermeneutics of the aggrandizement of rhetoric and the upsurge in the practice of law.

Chapter 2 focuses upon biblical hermeneutics from early Judaism through to Enlightenment. The chapter begins with an overview of a commonplace within contemporary continental philosophy of social science, which overlooks the importance of this Christian inheritance. We depict the main Ancient Jewish schools of interpretation and then discuss early Christian innovations through to the early church fathers like Augustine. Medieval practices are then depicted and both the reformation and later orthodoxy analysed.

The German philosophical developments during the eighteenth and nineteenth centuries are the subject of Chapter 3. We examine the changing face of interpretation with the rise in the desire for objectivity accompanying the intellectual climate of the growth of the sciences and reason in the Enlightenment. We concentrate upon the works of Wolff and Chladenius. We move on to witness the Romantic rebellion against this period and examine how the novel modifications of thinkers like Schleiermacher, von Humboldt, Droysen and Dilthey transformed hermeneutics into its recognisable modern form.

Chapter 4 explores the contributions of phenomenology and existentialism to interpretative theories and practices. We highlight the importance of figures like Husserl and Heidegger, and examine, too, the thought of the theologian Bultmann who added key ideas to this ever-important continental tradition.

Chapter 5 discusses the post-war hermeneutics of the twentieth century. We detail the major achievements of the pivotal figure of Gadamer. We show the debates and developments inspired by his work, beginning with the critique by Habermas, the innovations of the structuralist-inspired thinker Ricoeur and the attempted theoretical deconstructionist finale augmented by Derrida. We end with a detailed reflection upon the application of hermeneutics to the social sciences through an inquiry into the work of Geertz.

Part II explores the tradition of genealogy. Chapter 6 begins with an introduction to the history of genealogy and demonstrates how this approach was first articulated by Friedrich Nietzsche. We glimpse his life and works and early Romantic writings before describing genealogy. Then we show how his philosophical approach was developed in order to criticise Christian society. We portray the ontological and epistemological features of genealogy.

Chapter 7 narrates the deployment of Nietzsche's ideas as an approach to the humanities or social sciences. For this we turn to the work of the flamboyant Frenchman Michel Foucault. Most relevant to this subject matter are Foucault's later works. His essay is foundational on Nietzsche's genealogy named '*Nietzsche, Genealogy, History*', and important, too, are his studies of power and knowledge in the human sciences contained within his collection of works entitled *Power/Knowledge*.

Chapter 8 examines the main applications of genealogy in Foucault's work. Here we recount Foucault's own social sciences analysis and demonstrate in detail how it is an instance of genealogy. We study his genealogy of the penal system as disclosed in *Discipline and Punish* and then explore his work on sexuality in his three-volume study on the *History of Sexuality*.

Part III is dedicated to the tradition of Critical Theory. In Chapter 9, we profile critical theory from its historical roots. We start by drawing some links with the Ancients and then show how Christian ideas are influential to the founding fathers of the emergent eighteenth century 'critical rationalist' tradition – Kant and Hegel. We summarise the main ideas of these German intellectual giants and show how their influence continues in Marx and Later Marxism. We relate the trajectory of this inheritance to the Early Frankfurt School.

Chapter 10 surveys the formulation of critical theory itself in the work of Horkheimer, one of the principal members of the Early Frankfurt School. We introduce the first formal conceptualisation of the notion of a critical theory with Horkheimer's analysis of critical and traditional theory. We then move on to witness the application of critical theory in one of the Early Frankfurt School's key texts, the jointly authored *Dialectic of Enlightenment* by Horkheimer and Adorno. This represents a critical theory of their contemporary society in the form of a critical theory of the entirety of Western history.

Chapter 11 concludes our section on the tradition of critical theory. We survey two pivotal innovations in critical theory developed by other members of the Frankfurt School. First we look at the early Frankfurt School member, Adorno's unusual and remarkable *Negative Dialectics*. Secondly, we inquire into the work of the later Frankfurt School member, Jürgen Habermas and we assess this later contribution. We confine our analysis of Habermas to his important early period which is the bedrock in the development of his critical theory.

In conclusion, we summarise the central features of continental philosophy of social science. We draw together our three traditions and highlight how each one displays the main characteristics of humanism.

PART I

THE TRADITION OF HERMENEUTICS

H ermeneutics is one of the oldest traditions in the humanities. It pre-
dates modern scientific forms of enquiry and can be traced back
to ancient beginnings; it was pivotal to Greek education. As a practice, it
was highly influential throughout both Ancient and Christian times, and
because of these historical roots the term hermeneutics can, on occasion,
be used specifically to refer to the study of ancient texts and the Scriptures. Today, however, hermeneutics, although still predominant in the
old humanities disciplines of theology, classics, history, philosophy, law
and much of literature, has spread into modern disciplines like politics,
anthropology, sociology and even cultural studies where its techniques
are of vital use.

The term 'hermeneutics' is derived from the Greek '*hermeneutikos*'
which means 'to interpret', and it retains the same meaning today. Interpretation is a commonplace phenomenon and we could not with any
truthfulness depict an origin to this practice, be that historical, cultural
or geographical. Interpretation is something human beings throughout
time and history have undertaken. We have needed to interpret our world
in order to survive and also have wanted to understand its deeper meaning. In social and cultural interaction we interpret each other, often tac-
itly, on a day-to-day basis. When we talk of hermeneutics, however, we
mean something rather more specific. We are usually referring to a disciplined approach to interpretation often aimed at understanding texts of
special, cultural significance. Hermeneutics is mainly practised by scholars who have wanted to understand the meaning of important works
handed down to them from their ancestors. As such, interpretation is
an activity which has been involved in an attempt to gain knowledge,

communicate and establish the basis of authority for a canon of texts and agreed readings of them.

Like its counterparts in approaches to the study of the natural sciences, hermeneutics dwells upon issues of methodology and of foundational philosophical assumptions. However, its approach is very distinct from either that of the natural sciences or of the scientifically influenced social sciences. Indeed, hermeneutics is a practice that in many ways is so distinct that we could go so far as to say that it is the most extreme opposite to empirical (scientific) approaches to the study of the humanities. In fact, hermeneutics is a product of a specifically humanistic approach, one that has a deep trajectory within Western culture. We can characterise five main features that distinguish it from scientific counterparts. First, hermeneutics' main concern in contrast with the sciences is not with *explanation* but with *understanding*. For example, in the sciences we might want to explain a sequence of events that lead to the eruption of a volcano, or explain how a disease leads to certain symptoms. In hermeneutics, in contrast, we want to understand the moral, aesthetic, spiritual and historical significance of a work or cultural artefact.

A second and related difference to empirical science is that the former often deals with issues of causality, and attempts to subsume particular incidents under general laws. Questions of what causes volcanoes to erupt, the effects of viruses on the human body, or how chemicals react when mixed with other chemicals, are of interest to the geologist, medic and chemist. Hermeneutics in contrast is interested in understanding *meaning*. For example, the *Big Bang Theory* might give a scientific explanation of how we came to exist in causal terms, from various physical processes and reactions in space etc., but it offers no understanding about the meaning of human existence; why we have feelings, thoughts and moral conscience and how we ought to conduct our lives. Whether or not we are Christians and believe the Bible's explanation of the origins of man, we can see that this is clearly an attempt to give a different sort of account of the origins of human life: it deals with questions of human meaning and purpose in line with many other theological and philosophical texts written through the ages. We could say that science deals with 'knowing how', or, still referring to Ryle's categories, 'knowing that', and offers technical explanations, whereas hermeneutics is concerned with agendas of meaning and purpose.[1]

---

[1] Ryle 1990: 212–25.

A third feature of science is that it is *external* to the subjects it wishes to acquire knowledge about. A scientist is always outside the objects of his enquiry whether they be landmasses, cells or neutrons. Hermeneuticists on the other hand are within human culture and history. Consequently, they are often concerned with the *internal* aspect of experience, for instance, they might be concerned with what it felt like to be Napoleon, or with how people thought politically during the Renaissance.

Fourthly, scientists use external *methods* of observation and experimentation in order to gain knowledge. Hermeneuticists aspire to gain access to internal experience and to understand inherent meanings from within another culture – be it historically or geographically remote, or indeed, far removed in its traditions. They have a range of techniques, including linguistic, grammatical and historical skills. Their approaches range from more intuitive practises like the deployment of empathy and identification to just 'being'– where understanding is simply an inseparable part of what we are.

Finally, and poignantly, science often looks to the future for knowledge – we are seen to progress forwards in time to ever-greater sophistication in our awareness about the world. Hermeneutics, in contrast, often builds upon the past. Each new generation of scholars stands upon the shoulders of his or her predecessors, for, in its concern with interpretation, hermeneutics usually looks to the influence of previous readings of a work and holds a great reverence for the 'grand masters'. Indeed, we are often felt to be in the shadows of our forefathers. In the Classics, for example, we might read Plato, Aristotle and Cicero. In philosophy we would interpret the works of Descartes, Berkeley, Hume and Kant. Whilst in English literature, we would look to Shakespeare, Dryden, Johnson and Wordsworth. These canonical texts have acquired layers of significance through a history of interpretation and reinterpretation, each successive generation seeking to recover new insights from the past.

In addition to recovering texts from the past, hermeneutics itself as a practice has a deep history running from Ancient, Christian, Enlightenment and Romantic to contemporary times. Every historical era has seen a proliferation of theories about the nature of interpretation and indeed many competing modes of practice have arisen over the course of time. In fact, there are a plethora of varied and conflicting views about the best, and sometimes only, way to practice the art, beginning with ancient and medieval scholars and continuing into the present-day philosophical discussions. Because of its historical depth and complexity, it seems highly appropriate to view hermeneutics through a historical rather than conceptual perspective. However, this historical view, as we shall see, is not uncontroversial and is itself subject to many issues of interpretation.

From the standpoint of our own times it is quite clear that contemporary academic interest in hermeneutics is fairly ubiquitous across the humanities. This widespread appeal has been enhanced due to the popularity of recent influential, twentieth-century continental philosophers like Gadamer and Ricoeur. Moreover, a resurgence of interest in Heidegger has also added to the prominence of the discipline. In spite of its powerful status, however, any historical examination of hermeneutics leads to an abrupt encounter with an institutionalised division within academia. One side of this division allows accepted commonplaces about the role and history of hermeneutics to spread among a contemporary population. To witness this, we need to clarify just who practices hermeneutics today.

There are, very broadly speaking, two main groups approaching hermeneutics. First, there are those with contemporary interests; these may be entirely philosophical, students of Gadamer or Ricoeur for instance, or they may have practical goals, the implementation of hermeneutics in literary criticism, social science or indeed law, seeking to use and apply the skills of interpretation upon their chosen subject. Both these contemporary theoretical and applied hermeneuticists often seem to claim that a sophisticated awareness of issues in interpretation is almost entirely unique to our modern era. Warnke, for instance, in her otherwise subtle philosophical account writes: 'As a distinct discipline, hermeneutics had its origins in nineteenth-century attempts to formulate a theory of interpretation'.[2] Furthermore, she continues to claim that it was Schleiermacher, the Romantic hermeneuticist, who 'expanded the scope of hermeneutic questions' and addressed 'what the methods were that would permit an objective understanding of texts and utterances of any kind'.[3] In so far as hermeneutics has a deeper history than the nineteenth century, Warnke makes a general and passing reference to origins in a Greek educational system, but, as she expresses it, 'on Dilthey's account advances in the formulation of methods of interpretation had to await the reformation and the attack on the church's authority to interpret the bible'.[4] Prior to the reformation, Warnke suggests that interpretation was merely the act of repeating the accepted dogma of the Catholic Church.

[2] Warnke 1987: 1.
[3] Warnke 1987: 2.
[4] Warnke 1987: 5.

She cites Dilthey as claiming that modern hermeneutics represents 'a liberation of interpretation from dogma'.[5] Thus, we are left with the sense that hermeneutics proper only arises after the nineteenth century when the Romantic hermeneuticists Schleiermacher and Dilthey give it special attention, and whilst its more general origins may have been Ancient, it is only with the reformation that any real proto-history emerges.

The second, broad group of hermeneutic scholars challenge this assumption and approach the subject historically. Herein classicists and medievalists examine interpretation as part of the ancient curriculum of grammar or rhetoric. In contrast to the above contemporary approach, they point to how sophisticated forms of interpretative practice were a central part of ancient academic enterprise. For example, Copeland explains the detailed role of interpretation first, in the Greek grammarians' tradition and secondly within Roman rhetoric.[6] She further depicts the complex use of forms of interpretation within the early to later middle ages. Throughout, she emphasises the link between methods and issues of interpretation with translation. Kathy Eden also examines the historical depth of hermeneutics within the West.[7] She actually goes so far as to challenge the contemporary philosophical truism about the distinctly modern nature of hermeneutics when she writes that 'hermeneutics is not a German speciality but is grounded in the ancient tradition of rhetoric'. Indeed, far from having its most significant history in nineteenth-century German developments she argues strongly that hermeneutics is neither modern, nor indeed even a vernacular tradition but that it is grounded in Latin thought. Moreover, she rebukes Warnke's claim that perhaps a key historical moment, if one were to look for one, might be the reformation. For Eden the ancient Roman tradition of Rhetoric shapes hermeneutics and in view of this particular past, hermeneutics develops not through religious practice but from a specific legal and adversarial function.

In general, we see, therefore, that historical perspectives on hermeneutics fall into the two main camps. On the one hand, there are those that perceive it as a sophisticated practice and discipline arising only since what can roughly be described as 'modernity'. On the other hand, we have those who argue for a much deeper Western history. This conflict ought perhaps not to be expressed as one of complete opposition. What is at stake is not the denial of hermeneutics' past, but of its significance.

[5] Warnke 1987: 5.
[6] Copeland 1991.
[7] Eden 1997.

Warnke, for instance, does not refute hermeneutics' Ancient origins, nor does she deny that interpretation occurred before the Reformation. Her point is that these early modes were rather crude and unimportant in an assessment of the tradition. Copeland and Eden, in contrast, although disagreeing about exactly which historical moment and tradition was of central significance, do agree in so far as they believe wholeheartedly that these Ancient practices were highly sophisticated, both in relation to their own times – when they already included many of the features of supposedly modern hermeneutics – and as setting the foundations for later developments.

The problem that this fissure in the importance of the history of hermeneutics leaves is that, on the one hand, hermeneutics can be (wrongly) regarded as a principally modern 'invention' when its deeper past is denied. Contemporary hermeneuticists are then left with a depleted history of their subject and often a deeply mistaken view about the claims to novelty of recent developments. On the other hand, when the history of hermeneutics is recorded, it is often not analysed in any philosophical detail, and moreover, it is rarely placed in relation to later developments. This approach can sometimes mean that hermeneutics is treated as a historical relic: a museum piece. In the account which follows we try to fuse the merits of both these approaches. On the one hand, we seek to demonstrate the importance of the deeper historical past of hermeneutics. On the other, we hope that this will illuminate contemporary philosophical hermeneutics and its application to other disciplines.

Our overall account of hermeneutics is historical. First, we consider an historical account to be the most appropriate form of discussion of what is itself a historically sensitive tradition. Secondly, and relatedly, we wish to display hermeneutics' overall distinctness from scientific forms of enquiry. In the following chapters we depict the key developments in hermeneutics from Ancient Greek, Roman, Christian, Enlightenment and Romantic to contemporary philosophical contributors like Gadamer or Ricoeur. We examine the major contributors, depict their seminal works and analyse the key features in their debates about the nature and practice of formal, academic interpretation.

1

Ancient Hermeneutics

Depicting the Ancient Greek legacy to hermeneutics entails, of course, that we face the challenges of interpretation itself. There are no clear accounts of the origins of hermeneutics, that is to say, no historic moment when, as with Columbus's discovery of the Americas or Newton's moment of realisation about gravity, we can say hermeneutics was discovered, or indeed invented. Furthermore, to say even in the broadest of terms when hermeneutics first arises is to encounter enormous problems in the interpretation of ancient cultures. Gerald Bruns, for example, reflects these issues in his most interesting discussion in *Hermeneutics, Ancient and Modern*.[1] His broad point is that hermeneutics has multiple and conflicting histories, extending back before the origins of writing itself. He discusses the importance of Ancient Greece as a possible starting place for any perspective on hermeneutics' origins, but as he argues it is impossible to really trace back hermeneutic practice to a clear starting point. One of the reasons for Bruns's perspective is that he places little emphasis on the etymology of the term 'hermeneutics' but looks instead to an analysis of the act of interpretation as he believes it was performed by our ancient predecessors.[2]

In spite of the difficulties faced in defining the origins of hermeneutics, most scholars agree that the Ancient Greek legacy is a good starting place and some do perceive this as a kind of beginning. In view of this perceived

---

[1] Bruns 1992.
[2] Bruns 1992.

23

starting point Palmer, for example, focuses upon the importance of the etymology of the term 'hermeneutics' – from the Greek '*hermeneutikos*'.[3] He is, therefore, generous in his acknowledgement of our debt to the ancients and he devotes the opening chapter of his *Hermeneutics* to an analysis of the Greek contribution. Indeed, he begins his study with a claim that the beginnings of the word hermeneutics lie in this Greek verb *hermeneia*, generally translated as "to interpret".[4] Hermeneia, or interpretation, he points out, was discussed in several noteworthy accounts by the Ancient Greeks, from Xenophon, Plutarch, Euripides, Epicurus, Lucretius, and Longinus, to Plato and Aristotle. 'Aristotle', he writes, 'found the subject worthy of a major treatise in the *Organon*, the famous *Peri hermeneias*, '*On Interpretation*''.[5] Palmer explains that: 'the Greek word hermeios referred to the priest at the Delphic oracle. This word points back to the wing-footed messenger-god Hermes, from whose name the words are apparently derived'.[6] Indeed Palmer goes on to claim that the hermeneutic philosopher Heidegger accepted this Greek notion of hermeneutics as message bringing when he cited Socrates view (from Plato's dialogue the *Ion*) that the Poets themselves were, 'messengers of the Gods'.[7]

Furthermore, Palmer argues that Ancient practices of interpretation were not important merely as an origin but were also a sophisticated contribution to the discipline in and of themselves. According to his view ancient hermeneutics had specific features generic to all hermeneutics. For example, he believes that the Greek god Hermes was not only important because of the etymological link with hermeneutics but also because he represented many of the main aspects of all interpretative practice.

Hermes was not only important as a messenger but also was credited with the discovery of language and writing, and it was he who could transmute that which lay beyond human understanding into a form that the human mind could grasp. Palmer's point seems to be that Hermeneutics, for the Ancient Greeks, was a message-bringing process which entailed a 'bringing to understanding'.[8] Indeed, Palmer argues that this 'message bringing process of coming to understand' is implicit in all uses of

---

[3] Palmer 1969.
[4] Palmer 1969: 12.
[5] Aristotle 1853.
[6] Palmer 1969: 12–13. He adds, in parentheses (clearly making this association speculative), 'or vice versa'.
[7] Palmer 1969: 13.
[8] Palmer 1969: 13.

the term 'hermeneutics'. Thus, he makes three points. First, the origins of hermeneutics is in message-bringing. Second, the main feature of message-bringing is a coming to understand. Thirdly, all coming to understand is a kind of message-bringing. Although it seems to me that Palmer is rather muddled here as to whether message-bringing is a feature of understanding, or vice versa, his basic aligning of these two notions is an interesting point connecting the practice of understanding with a Greek icon.

Palmer goes on to conceptualise message-bringing in more detail. He develops three features. First, he claims, message-bringing involves '*expressing aloud*' (or the verb 'to say'); he offers the example of an oral recitation. Second, it entails *explanation* and third, it involves *translation*.

The first aspect of hermeneutics as entailing the verb 'to say', Palmer claims, is related to the 'announcing function of Hermes'.[9] 'The priest', he explains, 'at Delphi, brings fateful tidings from the divine. In his 'saying' or proclamation, he is, like Hermes, a 'go-between' from God to man'.[10] Palmer depicts four sub-features of *saying*. First, the way in which one expresses a meaning is important. Palmer describes this as the style of expression. Second, the ancient Greeks, he writes, often claimed to be inspired by the Gods in their 'proclamations' or expressions. Thus, inspiration becomes an essential component. Third, Palmer makes great emphasis of the oral nature of expression and argues for a magical power in speech, which the written word loses. Finally, Palmer claims that in expression, there is a temporal and productive element. He writes: 'the word must cease being word . . . and become event'.[11] Saying is performed over time.

A second overall feature of message-bringing according to Palmer is *explanation*. Palmer writes that 'words . . . do not merely say something . . . they explain something, rationalise it, make it clear'.[12] Palmer's depiction of explanation is, ironically, unclear. He describes explanation as first, explanation (a repetition which does not aid our understanding), second, as rationalising and third, as making something clear. He then offers the example of the oracle at Delphi which he himself describes as 'cryptic' and in 'words that concealed as much as they revealed'.[13] How then is explaining rationalising and making clear? He then moves on

9 Palmer 1969: 14.
10 Palmer 1969: 14–15.
11 Palmer 1969: 18.
12 Palmer 1969: 20.
13 Palmer 1969: 20.

in the same paragraph to add a further feature to his notion of explanation when he writes that 'explanation' is the 'saying something about
something else'. His point here seems to be that in saying, or expressing something, one is not merely *repeating* a prior meaning, but that one
moves onto a *new* meaning in the attempt to elucidate the original one.
Palmer's own analysis of Greek interpretation as containing a 'second
moment', a moment beyond saying and into explaining, is, according to
him, echoed in Greek thought itself. Palmer claims that Aristotle's *Peri
hermeneias* – a text which we have already mentioned as key to Greek analysis of hermeneutics – defines interpretation as 'enunciation'. Moreover,
'enunciation' Palmer concludes, after a long paragraph, 'is neither logic,
nor rhetoric, nor poetics, but is more fundamental; it is the enunciation
of the truth (or falsity) of a thing as statement'.[14] Interpretation conceptualised as enunciation in the Aristotelian fashion, is not concerned
with emotional or political effect – poetics and rhetoric, respectively –
but with truth or falsehood. In this, Palmer claims, Aristotle implies that
interpretation as enunciation, contains both the features of expression
and explanation. This is something of an unsatisfactory account, and does
little to clarify Palmer's notion of interpretation as explanation. His point
seems to be, however, that in the concern with truth or falsehood, one
moves from merely repeating a statement (as saying, or expressing aloud
would do) to attempting to analyse, and therefore, somehow explain or
rationalise it.

Palmer's characterisation of the third feature of Greek hermeneutics
understood as 'messenger-bringing' is that of *translation*. He claims that,
' "to interpret" means to "translate" '.[15] Later he writes that 'translation is a
special form of the basic interpretative process of 'bringing to understanding'.[16] Palmer argues that in translation one brings what is foreign into the
medium of one's own language. There is a problem with Palmer's analysis
as to whether he regards translation as a feature of all interpretation, or
whether he perceives it as a special kind of interpretation. Furthermore,
he goes on to add a further dimension of complexity by then claiming
that interpretation is *like* translation, and that a metaphorical relationship therefore exists between the two. Thus, Palmer's exact analysis of
the relationship between interpretation and translation becomes muddled during his analysis. However, the main point he seems to be making

---

[14] Palmer 1969: 21.
[15] Palmer 1969: 26.
[16] Palmer 1969: 27.

is that translation acts as a process of making the unfamiliar, familiar, and in that, it is an essential feature of the message-bearing process. 'Like the God Hermes, the translator mediates between one world and another'.[17] So translation brings forth the message from one language to another, and in that way is performing an interpretative role.

In short, the Ancient Greeks, according to Palmer, conceived of interpretation as a message-bringing process consisting in, first saying – expressing aloud the original meaning; second, explaining – transposing into a new form to elucidate the meaning; third, translating – making familiar. Palmer's claim that Hermes represents the feature of all hermeneutics is weakened by inconsistencies is his argument. First, it is unclear when he is offering an historical account and when he is developing his own speculative analysis of Greek influence. Secondly, he fails to depict the historical influence of Greek hermeneutics upon modern practice but merely describes similarities and parallels. Thirdly, the structure of his argument is muddled. Nevertheless, his analysis of the link between Hermes and generic hermeneutics is important and interesting.

There are other scholars who also give credit to Greece for the origins of hermeneutics. However, they place less emphasis on etymological issues and more upon Greek schools of scholarship and study. Copeland for instance argues that hermeneutics is indebted to the development of the Ancient Greek and Roman schools of *rhetoric* and *grammar*. She believes that grammar has a more significant role in the early formation of hermeneutics than rhetoric and thereby acknowledges, like Palmer, the debt to Ancient Greece, for it is only in Greece that the tradition of grammar is vaunted and broad. She also goes on to assess how Greek ideas about interpretation were carried forward through later humanist translation skills. Indeed, she argues, that more than the Greek origins themselves the development of hermeneutic practice owed much to the development of translation, from Greek to Latin and from the Ancient to the vernacular.[18]

Copeland certainly feels that the Ancient legacy is more than a mere archaeological ruin. For her ancient hermeneutics was important because the context of its practice contributed to the shaping of modern hermeneutics. Although she credits hermeneutics' central development to later medieval scholarly enterprises – she argues that it is in the art of *translation* that the major developments in hermeneutics occur, she

[17] Palmer 1969: 27.
[18] Copeland 1991.

acknowledges a substantial role for hermeneutics in Ancient times. She perceives that it is to the many twists and turns of the Ancient's oscillating emphasis upon philosophy, grammar or rhetoric that hermeneutics owes much of the nature of its early development. Within ancient Greece, philosophy was the highest attainment, higher than rhetoric. Grammar, lay ahead of rhetoric and had strong links with the Ancient Greek's concern with philosophy. Grammar itself was a broader and more important tradition. The ancient Greeks included within the tradition of grammar both exegesis and commentary. Therefore, grammar was not just concerned with technicalities of language but was far broader and included many issues of interpretation. In fact, for the Greeks (in contrast to the Romans) grammar was not kept too distinct from rhetoric so that issues of textual understanding merged with the art of persuasion. For instance, in Aristotle's *Poetics,* Copeland points out how Aristotle argued that interpreting poetry is related to the means of persuasion. Thus hermeneutic practice was transmitted through the Greek grammarians and owed many of its features to the grammarians' practice.[19]

Other scholars also look to Ancient Greece as an important starting point in the development that led to the formation of modern modes of interpretation. Mueller-Vollmer writes that 'besides the sacred hermeneutics of the Protestant reformers, three other tendencies were instrumental for the rise of modern hermeneutics: developments in classical philology, jurisprudence, and philosophy'.[20] With respect to the exact nature of this Ancient Greek influence, Mueller-Vollmer echoes the points made by both Palmer and Copeland. First, in accord with Palmer's view, Mueller-Vollmer distinguishes the features of ancient interpretation and pinpoints the etymological link between hermeneutics and Hermes. Like Palmer too, Mueller-Vollmer argues for the various roles and facets of interpretation revealed through this context: 'In order to deliver the messages of the gods, Hermes had to be conversant in their idiom as well as that of the mortals for whom the message was destined'.[21] He continues to explain that Hermes 'had to understand and interpret for himself what the gods wanted to convey before he could proceed to translate, articulate, and explicate their intention to mortals'.[22] And he concludes that 'looking at Hermes' task may give us a clear

---

[19] Copeland 1991.
[20] Mueller-Vollmer 1985: 2.
[21] Mueller-Vollmer 1985: 1–2.
[22] Mueller-Vollmer 1985: 1.

warning as to the complexities underlying the term hermeneutics and the hermeneutic enterprise itself'.[23] Still pursuing Ancient Greek references to the actual term hermeneutics, Mueller-Vollmer discusses Aristotle's *Peri hermeneias*.[24] In this text, Aristotle deals with the logic of statements and the grammatical structure by which subject and predicate combine in speech to reveal the nature of things. However, Mueller-Vollmer is careful to note that in antiquity the term hermeneutics occurred only sporadically and that the history of hermeneutics has to look beyond the merely etymological.

Like Copeland, Mueller-Vollmer also points out how Greek ideas were transmitted from Ancient through to modern times: it was because of the resurgence in the study of Ancient Greek texts which occurred during the Renaissance that Greek ideas were brought forwards into later European history. Moreover, still echoing Copeland he also highlights the later humanist interest in translating Ancient Greek into the vernacular and the concordant link between translation and interpretation. He explains: 'for some time, the theory of translation was subsumed under the category of interpretation, as for example, in works by the English humanist Laurentius Humphrey in the sixteenth century and by the French Bishop Huet in the seventeenth century'.[25]

Further scholars, although acknowledging the Ancient legacy to hermeneutics, look not to Greece but to Rome. Whilst Kathy Eden, for example, agrees that hermeneutics is grounded within a central ancient tradition, it is not to Greece that Eden ascribes hermeneutics, but to Rome. Although acknowledging the *background* to the developments of Roman rhetoric in earlier Greek forms, she views the primacy given to rhetoric by the Romans and the context within which that rhetoric developed as pivotal to the very formation of hermeneutics. It is to a discussion of the Roman legacy that we shall now turn.

ROMAN

A major line of argument in the historical literature on hermeneutics not only examines the Ancient Greek heritage, but looks to Rome, too. Scholars address questions like: How was the Greek practice of interpretation shaped and altered by Roman thinkers? How important was the

---

[23] Mueller-Vollmer 1985: 1–2.
[24] Aristotle 1853.
[25] See Mueller-Vollmer 1985: 3.

practice of interpretation to the Romans? How striking was the legacy of
Rome to the later formation of hermeneutics? Have any of the Roman
interpretative practices affected the way in which we interpret today?

When it comes to the influence of Roman ideas upon hermeneutics,
however, we find a radical divide in the literature. On the one hand,
there are those that perceive Roman rhetoric as the principal foundation of modern hermeneutics, for instance Eden.[26] On the other, there
are those who omit to mention the rhetorical tradition as even relevant –
Warnke would be a good example here. This conflict within hermeneutic
scholarship about the importance of Rome is striking, far more so than
the conflicting perspectives on the role of Ancient Greece – for herein
at least all scholars acknowledge that the Ancient Greeks are in some
way, however minimally, a source for modern hermeneutics. The Romans,
however, are perceived either to be at the very centre of the tradition or
not worthy of any mention at all! In many ways, this division is unsatisfying,
as there is little actual debate between hermeneuticists over the impact
of Roman rhetoric. For example, in their depictions of the major trends
within hermeneutics, many scholars (Palmer, Warnke and Ferguson, for
instance) merely *assume* that rhetoric played a minor role and, laying any
discussion aside, simply focus their attention upon whichever tradition of
thought they consider most appropriate to the history of hermeneutics:[27]
in some cases this is the biblical tradition, for instance Ferguson, or in others, it is modern German philosophy, for example, Warnke and Palmer.
There is, therefore, no debate – Roman rhetoric is simply ignored. No
wonder then the frustration of Latin scholars like Eden who believe –
and produce a sophisticated and detailed monograph to demonstrate –
that the Roman tradition of rhetoric is responsible not only for the extent
but also for the nature of the entire tradition of hermeneutic thought in
the West. Because of the lack of engagement of contrary arguments, we
shall detail here the main theses that depict Roman rhetoric as pivotal to
Western hermeneutics. The main figures here are Eden and Copeland,
each offering a quite different argument about exactly how the Roman
rhetorical tradition constructed and influenced hermeneutics.

Before examining Eden and Copeland's very persuasive arguments,
it is worth mentioning that not all hermeneutic scholars accord with a
polarised sense of the role of the Roman rhetorical tradition in the development of hermeneutics. Mueller-Vollmer, for example, perceives this

[26] Eden 1997.
[27] Ferguson 1986.

Roman tradition as part of the multiple and varied history of hermeneutics. He neither ignores nor credits Rome with a *central* position in hermeneutics' history.[28] What he does do is explore the strands of Roman interpretation carried forwards through later European intellectual history. He explains how 'the revival of interest in Roman law, which began during the so-called twelfth century renaissance in Italy with the concomitant efforts of scholars to elucidate the Code of Justinian (A.D. 533), led to the development of a special hermeneutics of jurisprudence. This special hermeneutics would soon spread across the Alps to the rest of Europe'.[29] The exact influence of this, he seems to claim, is principally to bring forth the marriage of the Roman legal tradition with interpretation in modern Europe.

Specifically, Mueller-Vollmer provides little detail of the exact nature of this union and its impact on European methods of interpretation. He does, however, explain that 'the rise and development of legal hermeneutics was intimately connected with the rise of philology'.[30] – although, this of course can position him as arguing that Greek thought plays a more vital role than Roman, for, as we shall see, the Romans subsume philology in service of the greater tradition of rhetoric. Mueller-Vollmer, however, claims that 'grammatical interpretation remained the basis of legal explication. In fact, grammatical interpretation remained an essential category for most writers in legal hermeneutics'.[31] His point seems to be that in the later European humanist traditions which directly preceded Enlightenment, Roman legal ideas and their link with interpretation were influential in the hermeneutic tradition. (However, both Copeland's and Eden's scholarship will show that it was the Greek tradition that linked interpretation with grammar rather than the Roman one.) Anyway, the later humanist scholars combined Roman emphasis upon law with the Greek grammatical tradition of hermeneutics. With this hybrid of ancient ideas, the humanists, generated a new form of hermeneutics. It was this grammatical interpretation that predominated jurisprudential hermeneutics and its relationship with other strands was, according to Mueller-Vollmer, thus:

the last writer in the humanist and Enlightenment tradition – in 1806 – defined the relationship between grammatical and other kinds of legal interpretation in

[28] Mueller-Vollmer 1985: 2–3.
[29] Mueller-Vollmer 1985: 3.
[30] Mueller-Vollmer 1985: 3.
[31] Mueller-Vollmer 1985: 3.

the following manner. Grammatical interpretation should be directed solely at the literal sense of a given law. It finds its limits only where the meaning of a law cannot be understood from ordinary linguistic usage. At this point, the "purpose" (*Absicht*) of the law and the intention of the lawgiver have to be considered 'logical interpretation'.[32]

Although Mueller-Vollmer's account of the impact of Roman law centres upon a later humanist revival of an amalgam of ancient ideas, and although he fails to discuss the Roman tradition in its original form, centred as it was upon rhetoric, he does at least register the import of Rome in the formation of hermeneutics. In the last analysis, however Mueller-Vollmer does take sides in the debates about the principally significant traditions in the formation of modern hermeneutics. Although acknowledging the important role of Roman interpretation to the discipline of legal hermeneutics, he nevertheless regards Rome's impact upon general hermeneutics as more discreet. He, therefore, limits his discussion of Rome to a very brief summary in the history of modern hermeneutics. Moreover, he concludes that it was not until modern German philosophy that we get 'a true watershed in the development of hermeneutics'.[33]

Kathy Eden's views could not contrast more strongly with thinkers like Mueller-Vollmer. She presents a dramatically distinct thesis about the importance of Rome. She argues that the entire tradition of Western hermeneutics is grounded upon and shaped by the discipline of Ancient Roman rhetoric. Consequently, according to Eden, if we wish to understand the art and practice of interpretation, we need not focus upon German philosophy or indeed the biblical tradition but instead on the history of Latin thought. Specifically, we need to examine the developments of Latin thought from ancient Rome through to the sixteenth century. The principal figures to read here are those that straddle the era from the Roman forefather Cicero to his sixteenth-century intellectual benefactor Montaigne.[34]

When we examine the voluminous works of these authors, we cannot, of course, escape the fact that rhetoric is bound up in the service and practice of law. If rhetoric is bound up with law, and hermeneutics is bound up with rhetoric, then hermeneutics' roots are embedded, Eden argues, within the tradition of the law courts.

---

[32] Mueller-Vollmer 1985: 3.
[33] Mueller-Vollmer 1985: 8 (this latter indeed forms the main emphasis for his collection of significant hermeneutic texts).
[34] Cicero 1942; Montaigne 1958, 2004. See also Eden 1997.

The tradition of the law courts has a very distinctive nature – quite unlike other intellectual arenas. One could even describe this as a 'mood', a sparring, adversarial one, antagonistic even, stretching across some sixteen centuries. Opponents gather across a divide each with a rapt audience before them, seeking to prove to an audience that their version of events is the correct one: that their *interpretation* is the truth.[35]

Interpretation has its origins in this muddy adversarial conflict, Eden persuades us, and this cannot but affect its nature. From the start interpretation is a competitive practice and one which must be conscious of its audience and its goals. Eden points to pivotal aspects of the role of hermeneutics in this Latin legal tradition. First, the tradition of rhetoric is used to *persuade*. In the law courts this persuasion occurs in an adversarial stance. Interpretation develops as part of the art of persuasion against one's adversaries. Second, she introduces a new point. Within Roman legal society, the Lawmaker is supreme. Consequently, it is the role of the judge in the law courts to be the living interpreter of the lawmaker. He must interpret the lawmaker's intentions to 'accommodate the infinite variety and variability of human circumstance to a fixed and generalised set of rules'.[36] Interpretation thus develops from assimilating the supreme law maker's will to various pragmatic sets of circumstances.

Eden underlines a third point. The art of hermeneutics arises out of Rome and develops through Latin thought. In claiming this, she, along with others of the historical 'school' of approach to hermeneutics, advocates that this tradition is not merely modern. However, in pinpointing the significance of the history of *Latin thought*, she purports that hermeneutics is not principally indebted to any vernacular tradition or practice. This latter point puts her at odds with her own closest rival for the argument of the centrality of Roman rhetoric to hermeneutics.

Copeland agrees with Eden about Rome's foundational role in the history of interpretation. However, Copeland presents a persuasive argument for the development of hermeneutics as subsumed within both the history of scholarship and the history of *translation*. First, in her history of scholarship, Copeland notes the two competing disciplines of grammar and rhetoric and their oscillating supremacy for the Greeks and Romans. This, she argues, has subsequent effects upon interpretative practice. Although thereby acknowledging the importance of rhetoric as well as interpretation in the development of hermeneutics for the Romans, she

---

[35] Eden 1997.
[36] Eden 1997.

does not, however, in contrast to Eden, focus specifically upon the Roman legal tradition as such.[37]

Secondly, her main thesis about hermeneutics occurs in her scholarly and detailed history of the development of translation, *Rhetoric, Hermeneutics and Translation in the Middle Ages*. It is a thesis that comes into play especially in her discussion of the advent of the vernacular and the subsequent need for translation and interpretation in the Middle ages. Her discussion of Ancient hermeneutics also includes reference to the importance of the Romans' own need for the translation of major texts from Greek into Roman. Thus, translation is a major player in the development of hermeneutics from Rome, both in the Roman's own translation of Greek into Latin, and in the later translation of Latin into the vernacular.

We have seen in our earlier discussion how the Ancient *Greeks* practiced the art of interpretation within the discipline of *grammar* and that philosophical concerns were uppermost. With the later advent of the dominance of Rome, a quite different emphasis was placed upon the role of grammar and it shrunk to a miserly range of technical issues and practices. As Seneca had it, 'the scholar busies himself with investigations into language'.[38] Meanwhile, in Rome it was *rhetoric* that grew to grand proportions. If the Greeks had exalted the truth, the Romans were concerned with power and influence. It was less the truth content of a statement that concerned them than how to persuade an audience that a statement was the truth. Hermeneutics therefore shifted from principally being preoccupied with issues of truth and unravelling this from the text, to issues of audience and effect. Cicero had it that rhetoric was a higher art than philosophy; Brutus declared that although intelligence was the glory of man, it was eloquence that was the lamp of that intelligence.[39] In *De Oratore*, Cicero declares that oratory is the summit of all human achievement.[40] Part of the thorough-going advancement of rhetoric was an expansion of its scholarly concerns. It was the rhetoricians, not the grammarians, who studied Greek and Latin historians and orators. Indeed, Copeland expresses it well when she writes that rhetoric expanded to deal with 'the text as a totality of audience, speaker, language, meaning and persuasion or effect'.[41] Rhetoric encompasses everything from the compositional inception – the speaker's or writer's invention – to the effect upon the

---

[37] Copeland 1991.
[38] Copeland 1991: 13.
[39] Cicero 1942: Brutus 15.59
[40] Cicero 1942.
[41] Copeland 1991: 14.

audience. She surmises 'rhetoric contains the most important questions of language, meaning and interpretation'.[42] The fact of interpretation now occurring within the rubric of rhetoric shaped its development: in Rome, until the fourth century, interpretation occurred within this tradition of rhetoric and issues of interpretation of meaning were inextricably bound up with questions about (audience) reception.

Hermeneutics also developed through a further strand of influence besides rhetoric. Rome, as we all know, was bilingual with Greek and Latin until the second and third centuries of Empire, then Greek became a foreign language. It was at this point that Greek became taught through basic translation techniques. Important Greek texts then had to be interpreted and translated to an increasingly distinct culture which viewed these texts as more and more alien. Interpretation in Roman times, therefore, combined with the discipline of *translation* which shaped it in quite distinctive directions. For example, in late antiquity, the praelectio – the introductory reading – situated the text and provided a formal introduction to the author. Exegesis itself was related to paraphrase. In late antiquity, significant developments, illustrated perhaps most usefully by Quintilian, displayed an increasing change in modes of interpretation.[43] Quintilian was principally interested in the Roman orators skills of rhetoric and these came to include an 'exact science' of interpretation. The text became a pre-given universal for which we could supply a fixed exposition.

Copeland's main interest, however, does not lie in the detail of Greek or Roman interpretative practices. Her interest in Roman hermeneutics is as a historical prologue to the later medieval practices. Her claim seems to be that it is in the Middle Ages through its association with the many facets of transition that translation undergoes, that hermeneutics gains its embellishment.

In short, we have four views about hermeneutics and the Roman legacy. Some scholars (such as Warnke and Palmer) entirely discount the relevance of Roman influence (a view which must itself be discounted as very short sighted), whilst others, perceive it as highly limited (Mueller-Vollmer). There are then those that regard the Romans as foundational in developing influential forms of hermeneutic practice. One of these perspectives (Eden) proclaims the absolute linkage between the Roman legal tradition and hermeneutics. A further view (Copeland), whilst acknowledging the role of rhetoric in the development of hermeneutics

[42] Copeland 1991: 14.
[43] Quintillian 2001.

believes that translation practices were more significant in shaping their interpretative counterparts. Whichever of these latter two views one takes, it is certain that rhetoric's supremacy in Rome influenced a line of thought and practice in hermeneutics whereby interpretation became concerned with issues of audience reception and the power of persuasion. This has been a lasting legacy in the West and can be seen arising in debates in modern philosophical hermeneutics, where some hermeneuticists identify the truth of the text entirely with the reader's, or audience's, view.[44]

Copeland's further thesis about the lasting relationship between translation and interpretation is one that merits serious study. Ideas raised through translators and the aspirations of gaining a correct translation, rage today throughout all disciplines of the humanities. For instance, a key issue is how to situate a text, whether to read it within its own historical context (and indeed how to gain access to this) or whether to appropriate the text for an understanding of the present. Contemporary historians such as Quentin Skinner, following a long tradition in British history, argue that texts ought to be read in the context of their own times. Others, especially those in more contemporary disciplines like politics, argue that one needs to understand the works of major thinkers like Machiavelli's '*Prince*' less in terms of what they tell us about the past, and more in view of how they can illuminate the present. These are just some examples of how the issues from the Roman rhetorical legacy continue into Western hermeneutical thought in the present day.

We have witnessed the debates about the importance or otherwise of the Ancient past in hermeneutics and also seen some of the key moments and ideas of this rich inheritance. Ancient Greek and Roman thought clearly represent one of the central pillars upon which all later Western hermeneutics stands. The other major pillar is an equally splendid one: that of the Biblical tradition. Thus we will go on to depict in Chapter II of our history of hermeneutics.

---

[44] In the contemporary subject of 'media studies' there is much discussion about the relationship between the truth and the power of persuasion of television. Questions addressed here implicitly echo Rome: has the interpretation of politics by the media merely generated a realm wherein the truth is what the audience is persuaded to think?

2

Biblical Hermeneutics

Views about the importance of the Judeo-Christian tradition per se and its significance to modern hermeneutics differ no less than opinion about the relevance and nature of the Ancient legacy. As with the Greek and Roman heritage, however, many contemporary philosophical hermeneutists are guilty of relegating the biblical tradition to a minor role. This oversight is accentuated by disciplinary boundaries and the fact that this strand of intellectual history is confined mainly to specialist theologians. Indeed, those from more contemporary disciplines like politics, sociology, anthropology, human geography, cultural studies and even law, often ignore the importance of two thousand years of Christianity altogether. Having said this, all philosophers at least would stand by the impact of the reformation upon hermeneutics. The remainder of the Judeo-Christian heritage, however, is fairly disregarded; indeed, certain philosophical hermeneutic accounts overlook the important developments that occurred in this tradition altogether. Warnke, for instance, hardly mentions the Jewish tradition or the many distinct Catholic practices of hermeneutics. She collapses the whole biblical tradition prior to the reformation into a single notion. Her assumption is that biblical interpretation prior to the arrival of Protestantism was a uniform practice. This argument is worth depicting because unfortunately it represents a commonplace among many contemporary hermeneuticists. Her thesis is roughly as follows.

During the Medieval Christian era, Warnke's argument goes, hermeneutics was the common practice of scholars who were concerned with interpreting the most important text known to human beings at this time, namely, the Bible. The first feature that was distinctive about reading the Bible during these times, Warnke alleges, was that there was an agreed 'correct' reading. There was only one Church, the Catholic Church, and according to the doctrine of the Church, there was only one accepted version of the Bible. Thus, the practice of hermeneutics was a practice of reproducing a single, *correct* reading. Note, therefore, that in this medieval system of interpretation, there is, according to this thesis, no place for competing interpretations or argument.[1]

The second feature of these early Catholic readings of the Bible, Warnke points out, was to reveal a true meaning. Hermeneutics was, therefore, a practice of unveiling the *truth content* of a text. The truth content of these Catholic readings of the Bible had two components. On the one hand, it consisted of a *metaphysical* dimension – an agreed 'truth' about how the world was created, the notion of God, questions about the existence of evil, the role of Christ and so on. On the other hand, the 'truth' content had a *normative* component, that is, a moral dimension. In the early biblical readings of the Bible, however, *moral facts* were as central as empirical and metaphysical facts and held the same epistemological status. This early 'doctrinaire' way of interpreting the Bible meant that the *best* reading was in fact the one which most accurately captured the view of the Church and was also in every sense, the *most conventional*, the *least original*.

With the arrival of the Reformation, Warnke argues, a challenge to hermeneutics was underway. With the Reformation came the advent of Protestantism, and in the wake of the tide of this new creed of Christianity came a *novel* reading of the Bible. However, although the Protestants offered a distinct interpretation of the Bible from the Catholics, Warnke claims that the actual approach to reading was exactly the same as the old Catholic approach, in the sense that their interpretation was again a form of doctrine. Protestants, like Catholics, wanted to know the truth content of the Bible. They wanted, first, the metaphysical truth about the nature of creation, God, evil, the role of Christ and so on. They also derived from the Bible a normative system of rules about how we should live our lives. Thus, in spite of the very different readings of the Bible provided by these rival Churches, both pursued exactly the same hermeneutic 'method'.[2]

Taking aside the similarity in the way of reading practiced by these two Churches, a great new development in hermeneutics had, Warnke

---

[1] Warnke 1987: 5–6.
[2] Warnke 1987: 5–6.

believes, in fact emerged. Instead of merely one correct reading of the Bible, there now existed two *competing* readings; each one claiming to be 'correct'. Moreover, one reading, the Protestant one, criticised the Catholic one as being flawed. As a result of this, Warnke continues, there emerged discussion about which was the correct reading. And in turn, as a consequence of this, people began to become conscious of the actual *process* of reading. Questions began to be addressed such as: How should one read a text? How should one arrive at the single, true meaning? What were the 'rules' by which the truth content of a text could be discovered? How could we know which reading was the correct one? These questions meant that the art of *interpretation* had developed. One text could contain two very different meanings. In fact, with the advent of Protestantism, came not simply one new religious doctrine, but many – multiple forms of Protestantism arose. And each of these forms had a distinct interpretation of the Bible. There therefore arose the notion that a single text could contain multiple readings.

The idea of this transition to a text containing several meanings is, of course, Warnke claims, one that *we today* can attribute to the Reformation. At the time, however, each brand of Christianity advocated its own reading as the true one: they did not regard a multiplicity of readings as legitimate. What did occur though, during the reformation, was the advent of discussions about how to arrive at these 'correct' readings and defend them against other 'incorrect' interpretations. So discussions arose about the *practice* of interpretation – that is, the practice of hermeneutics. According to Warnke's account, the vast majority of biblical hermeneutics is barely interpretation at all because it is merely the practice of reproducing a single correct orthodox account of the Bible as sanctioned by the Church. Even when later Protestant accounts challenge the Catholic one, they still follow the same technique she alleges, the dogmatic reproduction of orthodoxy.[3]

Warnke's points about the reformation are not entirely without merit, but her narrow idea that all biblical hermeneutics can be conceived as the pursuit of dogma is vastly overstated. Moreover, her attempt to summarise such a huge swathe of history and tradition in a few sentences does her little credit. Although some contemporary accounts of hermeneutics do simply ignore the biblical tradition or simplify it in the manner of Warnke's account, it is important to remember that theology and relatedly scriptural interpretation are traditions steeped in Ancient and historical

[3] Warnke 1987: 5–6.

scholarship and it would be unlikely for them to display the crudity that Warnke alleges. Moreover, as we shall see in what follows, the reformers were only one part of a much broader and sophisticated range of hermeneutic ideas and practices pursued by Ancient Jews, early Christians and medievalists as well as those of the later reformation. Having depicted Warnke's thesis about Christian hermeneutics as representative of a common misconception, we will now provide a more detailed history of some of the major competing ideas and practices of the biblical tradition. It is important, even for those with contemporary interests, to be aware of the biblical tradition as part of the West's intellectual legacy; at the very least, it guards against any simplistic claims to novelty based upon an ignorance of the past. For instance, some of the key issues, which later German philosophical accounts discuss and debate, are present in earlier Judeo-Christian interpretative practices.

In what follows, we depict the main tenets of biblical hermeneutics from Judaism and early Christianity through to the early church fathers (like Augustine), medieval practices, the reformation and later orthodoxy. In so doing, we use the term 'biblical hermeneutics' to refer to the activity of interpretation across an immense historical span. This stretches from Ancient to contemporary times. Ancient Jewish hermeneutic practices occurred after the Greek impetus and against the backdrop of the dominant Greco-Roman tradition. Although there are some overlaps in the nature of hermeneutic practice between Greco-Roman and Jewish cultures, the main line of influence is from Ancient Greek to Jewish practice. The later Christian tradition, stemming as it does from Jewish descent, follows on from this. We depict the biblical tradition until the seventeenth and eighteenth centuries when it is disrupted by the new scientific rationale of the early moderns and enlightenment. At this point, biblical hermeneutics is embellished by the emergence of the new philosophical tradition which appropriates a concern with interpretative issues. Although all aspects of biblical hermeneutics continue to develop beyond this time, many of the changes either reflect, or are derived from applications and appropriations of philosophical positions, and so are included in our later philosophical discussion. In short, by biblical hermeneutics here we focus upon the era from the Judaism of the Greco-Roman period to the seventeenth century.

Beginning our discussion with early Jewish hermeneutic practices, we see the dawning of a new focal point: the primary concern becomes the Bible. A far cry from the many literary, historical and legal texts of antiquity, interpretation now centres upon this one sacred text. The mainstay of Judaism, and as we all know the common point for all biblical hermeneutics, was of course the Scriptures. Although Jewish and Christian interpreters had interests in poetry, historical and legal writings among other things, it was an understanding of the Scripture that was clearly pivotal. These texts were known by the Jews as the Torah and they revealed God's (Jahweh's) own words. Moreover, they contained the complete truth for human life. Jewish hermeneutics reflected some of the divisions and debates of Greek and Roman interpretation.[4]

In the Jewish tradition, there were four basic schools of approach. First, *literalist* schools took the Scriptures at 'face value' and were mainly concerned with moral and legal rules. Second, Midrashic, rabbinic schools, led by Rabbi Hillel (first century B.C.) formed a specific set of seven rules governing interpretation. They emphasised the importance of historical context and thus generated a broader interpretation than the literalist schools. Their rules were strict in an attempt to safeguard against what they perceived as possible misuse of the Scriptures. The third school was rather specific, the Pesher School which sprang from the Qumran community and claimed special knowledge of the divine mysteries. This latter went so far as to relate biblical prophecy to their own contemporary events. The final school developed from Philo of Alexandria (25 B.C.– 50 A.D.) and represented the typically Alexandrian mode of *allegorical* interpretation. Herein the text was to be treated symbolically, to be interpreted at depth and its hidden meanings uncovered. The aim of this allegorical school was to show the Greco-Roman world the 'rationality' of Jewish Scripture.[5]

Already, we see in the Ancient Jewish tradition a multiplicity of competing interpretative practices. Against Warnke and other contemporary hermeneutical accounts, early Jewish biblical hermeneutics was much more than the reproduction of mere dogma. Each school developed a set of guidelines and rules about interpretation, and understood the role of the meaning of a text in very different ways.

Early Christian hermeneutics reflected many of the concerns of the newly developing Church. Several factors were crucial in defining this. First, against Judaism, the event of Jesus, his proclamation, death and resurrection were crucial. As a consequence of this, a shift occurred within hermeneutics which involved its very role. Because of the advent of Jesus, early Christianity underwent a partial de-centring of the role of the

---

[4] See Ferguson 1986: Ch. 9 and Jeanrond 1991: 12.
[5] Jeanrond 1991: 14–22.

scriptural *texts* themselves. For Christians it was the *experience* of Jesus Christ as embodying the fulfilment of God's promises that became key and this engendered the emergence of an extra-textual perspective and a strand of ideas which emphasised a practical response to God's presence. Hermeneutical concerns and debates took a less central place than in Ancient Judaism.

Secondly, when it did come to hermeneutical issues, Christianity saw the emergence of a completely new set of scriptures and these required interpretation for the first time. Moreover, they re-interpreted the old scriptures. To accomplish this latter task, Christians in fact appropriated the Hebrew texts and interpreted them narrowly according to the life of Christ. However, they used very similar methods of interpretation to the Jews. In fact, Jews and Christians alike practised hermeneutics of a *literalist* or *allegorical* kind. Two Christian schools of interpretation emerged, both based upon their Jewish forebears. First there was the *Alexandrian School*, which remained allegorical. In Alexandria the Christians were influenced by Plato. *Origen*, circa 185–254, who was the most prolific of the Christian writers of his time, assumed the leadership of Alexandria's Catechetical School in 203 at the age of only eighteen. In his major treatise, *On First Principles*, he showed how theology shaped hermeneutics.[6] Hermeneutics was to be the disclosure of the text's spiritual sense. Moreover, scripture contained the ultimate mystery which could only be understood in symbols so that allegorical interpretation was foundational.[7]

Secondly, the School of *Antioch* produced grammatical and historical accounts. Herein they looked for the historical reality within scripture rather than any hidden meaning. The true canon had to make reference to Christ's historical presence, and methods of textual and grammatical interpretation helped unveil the historical reality. Thus, in early Christian hermeneutics distinct approaches with their own special techniques of interpretation arose.

True Christianity, however, was not provable on hermeneutic grounds alone. Whether Alexandrian, or from Antioch, an interpretation was only valid if the interpreter aligned with the Church (perhaps this is the point which Warnke wanted to emphasise). However, although all Christian tradition held this viewpoint, some were more vehement than others. The Apostolic tradition was associated with *Irenaeus* (c. 115–c. 202) who was

[6] Origen 1985.
[7] Ferguson 1986: Ch. 9 and Jeanrond 1991: 14–22.

born in Smyrna in Asia Minor, where he studied under Bishop Polycarp (who in turn had been a disciple of Apostle John). This tradition held that an interpretation was only valid if a community of Christians were able to provide a framework to guide the account.[8] (Origen accepted this point and his interpretations were guided by the Church.) The *Tertullian School* meanwhile was far more strict and maintained that scripture was the property of the Church and that no interpretation was even possible outside of the Church. Indeed, there existed no legal right to interpret the Scriptures in any circumstances beyond the confines of the church itself (a point challenged by the Gnostics).[9]

We can see a number of issues arising in Jewish and Early Christian hermeneutics. First, between the two religious traditions we see a disagreement about how central the role of the religious texts are. Secondly, within each religion there are varied views about the nature of meaning in the text, whether it is literal, in the sense of historical, legal or moral, or whether meaning is allegorical and spiritual. Furthermore, different schools place differing degrees of emphasis upon the nature of the interpreter himself. Whilst all schools involve the notion of an agreed community of interpretation, be it a sect or the gathered Church, they vary in their views upon the degree of control this community should exert upon the interpretative act. Warnke's narrow appraisal of Judeo-Christian hermeneutics is clearly rebuked, even with a consideration of simply its earliest phase.

Augustine (who lived from 354–430) wrote the seminal hermeneutic text, *De Doctrina Christiana*, which brought a fascinating new insight into interpretative practices.[10] He rejected the literalist readings of scripture and moved towards the allegorical tradition. In so doing he accepted the need for rigorous interpretative method to prevent arbitrary readings, and drew upon his thoroughgoing classical education in rhetoric to develop a linguistically sensitive approach. In fact, Augustine looked to the significatory aspect of language and was the first scholar to develop semiotics. Semiotics should be understood as the development of the study of signs and symbols especially in language. (Note that hermeneutic scholars of a contemporary disposition not only fail to mention that many distinct interpretative practices occurred in earlier traditions, but

[8] Irenaeus 1920.
[9] See Ferguson 1986: 135 and Jeanrond 1991: 14–22.
[10] Augustine 1995.

they almost always fail too to accredit the 'invention' of semiotics to Augustine and usually accord this to much later academic developments.)[11]

Augustine's interpretative approach developed a number of distinct features. First, the semiotic view of language meant that Augustine was against a literal reading of texts. Secondly, from semiotics, the understanding of the nature of the sign is that a sign is not identical with what it refers to. From this, Augustine arrived at the notion that language understood as a signifier signifies something beyond itself. That is to say, in the case of scripture, the language of the Bible refers to God but is not itself identical with God. Augustine considered the Scriptures not as the word of God, but as *human* texts that referred to God. This latter raises an important point. In biblical hermeneutics, the issue of authorship arises. Views differ about whether the Scriptures are the actual word of God or whether they are human texts that refer to, that is to say, signify God. Earlier Christian thinkers believed scripture was the word of God: Augustine challenges this and makes humans their author.

In assigning a human authorship to the scriptures, Augustine reassesses their role. They are human texts which guide us towards God but are not themselves the presence or voice of God. That is to say, they are a *means* to reaching God. They are not the *end* of God himself. In order to pursue the means to the end of reaching God, we need to deploy semiotic methods in textual interpretation of the Scripture. Moreover, we need allegorical interpretative methods because a literal reading would entail that we mistook the means for the ends. For Augustine, although the means and not the end, scripture is no less important. It is still through texts that we reach God.

Augustine embellishes earlier Christian hermeneutics with an important and original further point. He develops a notion of the state of mind of the interpreter. According to Augustine, the interpreter must read with an attitude of 'love'. 'The love of God and of our fellow human beings is the proper reading perspective for a Christian believer'.[12] Love derives itself from scripture so that Augustine's notion of reading with an attitude of love, is circular. Love is the basis from which to approach scripture; scripture instructs the Christian community how to live according to love; the community living in accordance with this love forms the basis for reading the Scripture and so forth. (Note that this is an early precursor of the modern notion of the hermeneutic circle.)

[11] Jeanrond 1991 gives an excellent summary: 22–26.
[12] Augustine 1995; Jeanrond 1991: 23.

Far from Warnke's idea that biblical hermeneutics prior to Reformation is a singular tradition merely repeating Church orthodoxy, we see in Augustine a sophisticated awareness of many hermeneutical concerns. Not only does he address issues of how important texts are, the nature of meaning and the role of the interpreter, but he addresses issues of authorship, the state of mind of interpreters and indeed develops a semiotic approach to linguistic interpretation.

Augustine continued to be enormously influential in the Early Medieval era. For instance, Hugh of St. Victor's work *Didascalicon* was heavily influenced by *De Doctrina Christiana*, and consisted in an allegorical interpretation of scripture.[13] However, later critics of Augustine sometimes accused him of biblicism because of his intense focus upon reading scripture as the main foundation of Christian faith. Eventually, Augustine's hermeneutics was superseded by a new approach to biblical interpretation in the medieval era. This inaugurated two shifts. First, it changed the focus from reading texts to the importance of the Church and its institutions. Second, whilst Augustine had seen biblical interpretation as the basis of theological understanding, this was challenged by a new approach which separated scripture from theological concerns.[14]

In the twelfth century, with the rediscovery of Aristotle's texts, a new literalist approach to hermeneutics emerged and the interest in allegorical interpretation of scripture declined. Thomas Aquinas, the greatest medieval theologian, at the beginning of his *Summa Theologica* became involved in a discussion of hermeneutics.[15] His scholastic theology was representative of the new trend. Literal interpretation became the norm and hermeneutics evolved into an immensely detailed scholastic focus upon the literal truth of a text. Moreover, the discussion of deeper meanings within the texts that had been so precious to Augustine, continued its shift into the realm of theological speculation.[16]

Thus, the overall tenet of medieval hermeneutics was the appeal to the authority of the Church for the correct reading, an etiolation in scholastic literalism whilst the deeper spiritual meanings so precious to Augustine, became marginalized from interpretation, and shifted to the newly developing domain of speculative theology. All this could be grist

---

[13] Hugh of St. Victor in Illich 1993.
[14] See Jeanrond 1991: 22–26.
[15] Aquinus 1981.
[16] Jeanrond 1991: 26–30.

to the mill for Warnke, except that we must remember that this literalist hermeneutical approach derives from only one particular historical strand of hermeneutics, and although rising to become dominant for an era, is not representative of a singular biblical approach.

During the medieval era the institution of the Church dominated both theology and biblical hermeneutics. However, in the late fifteenth to early sixteenth century all this changed. A powerful challenge to the medieval Church arrived with the Reformation. This was a many faceted phenomenon, but for hermeneutics it entailed several interesting components. First, texts began to appear not only in Latin but in the *vernacular* – Copeland points to this as being hugely significant for hermeneutics. Secondly, there occurred a major shift in authority away from the Church as pivotal to the new underlining of the importance of *faith*. Indeed, the emphasis was upon *individual* faith and individuals reading scripture. Thirdly, against Augustine's earlier interpretative approach, the Bible was perceived to be the word of God and *literal* interpretation thus continued to dominate against its allegorical counterpart.

Of the greatest minds of the reformation, Luther and Calvin stand out. Both held that 'the Holy Spirit was the 'author' of the scriptures, and that, therefore, God does speak through the texts of the Bible in a way that was clear, coherent and sufficient for human salvation'.[17] Whilst Luther developed a rather specific christological interpretation of scripture, Calvin, who was a more devoted humanist, perceived the Bible itself to be the ultimate basis of Christian faith rather than a particular reading of it.[18]

We can see through Calvin the broad tenets of protestant humanist hermeneutics. Against the early Church fathers, Calvin made *scripture*, not the tradition and authority of the Roman Church, central to Christianity. Reading the Bible was the basis of a Christian's religious life and scripture should be read through grace and faith. Indeed, an *individual*'s grace and faith were more important than blindly following the authority of the Church. Calvin believed that this grace and faith entailed spiritually disciplining the soul to prepare it to be completely receptive to the works of God. And because the word of God has to be preached and received with the whole soul, this discipline entailed a critical, indeed almost scientific preparation.[19]

---

[17] See for example Calvin 1845; Luther 1991 and Jeanrond 1991: 31.
[18] See Uffenheimer, B. and Reventlow, H. G., eds. 1988.
[19] Calvin 1845; see also Torrance 1988.

Although shifting the emphasis from Church and tradition to individual faith, Calvin was not advocating a subjective mode of interpretation. Indeed he was entirely against a subjective component in interpretation. However, having relinquished the absolute authority of the Church (this too as we shall see could be subjective) Calvin needed to find a new source of objectivity in faith. He found this through emphasising the need for an *objective interpretation* of scripture.[20]

This objectivity in reading the Bible was obtained through an attempt at knowing oneself. The interpreter had to know himself in order to diminish his subjective component.[21] Individuals had to know their own prejudices and presuppositions in order to attempt to remove them. Then, with a disciplined spirit they could submit to the word of God. Calvin's point was that reading would be animated by the will of God itself.

Calvin's emphasis upon an objective reading of scripture spread to the Church itself. He was convinced that the Church had to gain critical self understanding. Indeed, Calvin was as much against any subjectivity in the church as he was hostile to it in the individual. The whole church must itself attempt to critically discipline itself in order to attain the desired objectivity in its interpretation of the Bible: it too must submit to the word of God.[22]

The important shift from tradition to disciplined faith as the basis for interpretation cannot be over-emphasised. However, it should be noted that Calvin allowed for some role for the Church, for tradition, and for the early Christian fathers as a basis for interpretation.

In contrast to Augustinian views, for Calvin the Bible was the actual word of God. In fact, scripture was God's word telling of God's work on man in history. As such, the Bible contained the ultimate truth and interpretation was pivotal as the means to understanding truth. In view of the awe for this truth, Calvin was firmly against the use of rhetoric and equally hostile to novelty or ambition in biblical exegesis. However, in spite of holding firm beliefs about the nature of the correct procedure for interpretation Calvin thought that the true Christian ought not to get involved in hermeneutic debates as these would present a distraction from the main purpose of gleaning the Bible's meaning. In a similar vein, Calvin was also against the application of external theories in

---

[20] Calvin 1845.
[21] I say *he* as we were at this point in history mainly referring to men.
[22] Calvin 1845; see also Torrance 1988 for a detailed account of Calvin's hermeneutics.

interpretation. He believed that one should not move beyond the internal structure of the text itself. In this way he was opposing the trend of medieval scholarship and, in common with Augustine, was reverting to the text as itself central. However, unlike both the medieval and early Church fathers, Calvin rejected both sophisticated theories and techniques of interpretation and proposed an effective and dogmatic approach to interpretation. After all, the Bible was principally a guide for how we should live our lives as true Christians.[23]

In terms of the tradition of hermeneutics, it is hard to see if the Reformation in general, or if either Luther or Calvin in particular, contributed much that was new to the practice of interpretation. Ironically, although the Reformation developed a radically new interpretation of Christianity, its methods of textual interpretation remained closely aligned to the early Antiochian School of literal interpretation. The emphasis upon scripture as central echoes Augustine, and the emphasis upon reading in a particular state of mind, namely that of grace and faith, were not new – Augustine had already focussed upon reading in a state of love. However, there was perhaps one important (dare we say) novelty to the Reformers' hermeneutics. Having removed the centrality of the Church as the basis for the authority of a reading, they were wide open to individual, arbitrary interpretations. To guard against this, they had developed the notion of the importance of *objectivity*, and as such, the Reformers paved the way for a later discussion of how objective interpretation of a text could be attained. The Reformers aspired to an almost scientific method of reading and thus prefigured seventeenth- and eighteenth-century preoccupations. Indeed, they paved the way to the central concern of Enlightenment hermeneutics.

Before moving on to assess the significant development of Enlightenment itself, we should, however briefly, mention the hermeneutics of Roman Catholic and Protestant orthodoxy. Roman Catholic theologians gathered in the sixteenth century and confirmed their belief – against the Protestants – in the Christian church and the Bible taken together, as the basis for true belief. In response, later Protestants took a more extreme stance than their Reforming founders had done, and resorted to the maxim of biblical infallibility. The Bible was dictated by the Holy Spirit and as such, anyone who doubted the literal truthfulness of any part of the Bible was not a true believer. Herein, a dogmatic set of readings said to be the true readings were presented by protestants as alone

[23] Torrance 1988.

representing the orthodox faith. Critical and individual readings were denied and the Bible was reduced to dogma. In fact, both Catholic and protestant orthodoxies developed, each with its own dogma, each firmly rejecting the other. Moreover, these orthodoxies were antagonistic to the developing rationality and science of the Modern era.[24]

However, reason and science were moving inexorably forwards, and developments in philosophy, notably Descartes, meant that the theologians in the seventeenth century began to accept a separation between philosophy and theology. Scripture was no longer the only source of knowledge, and indeed science could reveal truths and had methodologies that were in some realms, superior to theology. Thus, hermeneutic concerns were challenged in a new way. Not only was the interpretation of one set of works an issue, and not only were multiple texts being subjected to hermeneutic practices, but also issues about method in interpretation were to be raised to the fore.

[24] See Jeanrond 1991: 35–39. Not that the hermeneutics of Roman Catholic and Protestant Orthodoxy do in fact represent Warnke's views. However, as we have seen, these are but one minor strand of a much more complex tradition.

3

German Philosophical Hermeneutics

Enlightenment and Romanticism

ENLIGHTENMENT AND ROMANTICISM

In moving on to assess the hermeneutic contribution of the dawning modern era, we are entering the realm of German philosophical hermeneutics. In spite of making a strong claim throughout our discussion thus far for the significance of earlier Ancient and Judeo-Christian interpretative practices, this is not in any sense meant to diminish the value of the German contribution. As Warnke, Mueller-Vollmer and Palmer have all so well told, this is undoubtedly a mighty tradition. My point is that this later tradition builds upon earlier humanist strands of thought. Ironically, a further point is that those humanist scholars with their distinguished accounts of Ancient Greco-Roman and Judeo-Christian hermeneutics, for instance, Copeland, Eden and so on, can in turn underplay the significance of the later German contribution.

Philosophical hermeneutics grew on the continent. It can be conceptualised as beginning roughly around Enlightenment and then developing through Romanticism to the twentieth century and beyond. Its main protagonists during the era of Enlightenment were Wolff and Chladenius. Throughout the Romantic period, Schleiermacher, von Humboldt, Droysen and Dilthey all carried the flame. Later, phenomenology and existentialism, with figures like Husserl and Heidegger, and the theologian Bultmann became key. Post-war hermeneutics of the twentieth century, meanwhile, has been dominated by Gadamer, with later thinkers like Ricoeur and Derrida initiating important developments and, of course, Habermas intercepting with influential debates.

THE IMPACT OF ENLIGHTENMENT

We begin our discussion of continental hermeneutics with the advent of Enlightenment. With Enlightenment there came many shifts in the focus of knowledge and these impacted greatly upon the practice of hermeneutics. The eighteenth-century German philosopher and grand 'luminaire' Immanuel Kant, in his extraordinarily influential essay, notoriously captured the ethos of this era. He addressed a question that in a variety of ways was to have an enormous influence on the practice of hermeneutics. The essay of which we speak is of course Kant's '*Was ist Aufklarung?*', translated into the English as '*What is Enlightenment?*'. In this essay, Kant argues that man emerges from his self-incurred immaturity by using his own reason rather than merely following doctrine.[1] Foregrounding the importance of reason impacted upon the hermeneutic tradition in at least three ways. First, the assumption of Enlightenment, as in previous biblically orientated schools of interpretation, was that a text was assumed to have '*truth content*'. This differs from the earlier Roman tradition of rhetoric wherein greater emphasis was placed upon the importance of the reception of texts, and, as Eden was so keen to point out, on interpretation as aiming towards influencing the law courts. In Enlightenment, moreover, in contrast to Christian traditions, *reason* was the means to gain access to this truth.

Secondly, following on from reformation hermeneutics, the trend away from Church authority to '*objectivity*' was reinstated in enlightenment. The Enlightenment's advocacy of reason as a means to discover truth challenged dependency upon the Church's authority. Herein we encounter the simplified view of the Enlightenment as a whole: reason-challenged doctrine. Although there are complex relations between reason and religious authority (many key Enlightenment figures were also Christians), it is fair to say that the introduction of reason as the grounds for the legitimacy of the truth was the most significant cultural and intellectual transformation in Enlightenment.

For hermeneutics the refusal of the mere assumption of the Church's authority and the advocacy of reason introduced questions about the grounds of the legitimacy of the truth content of the text. In promoting dependency upon reason instead of 'authority', Kant had generated a shift towards reasoning about the origins, authenticity and legitimacy

---

[1] Kant 1977. Importantly, Kant as we all know, was not promoting an atheistic argument, but this issue remains material for another discussion.

of knowledge. Kant's own question, roughly paraphrased from his great Critique, was: 'what are the grounds for the legitimacy of our knowledge?'[2] Hermeneuticists pursued the same question and adjusted it to their own concerns. They asked: 'what are the grounds for the legitimacy of an interpretation?.'[3] Thus, one could no longer merely be told by a canon what a text meant, but one needed to know *why* that reading was valid.

Thirdly, Enlightenment's association with science and logic and the related dissemination of rational systems of understanding in all branches of the humanities affected hermeneutics. Interpretation was believed to be governed, like all other sciences and like logic itself, by universal rules. Interpretation, which throughout Western history from later antiquity through to the seventeenth century, had been within the province of the humanities, now became a central topic of analysis specifically within philosophy. Following Aristotle, Enlightenment philosophers viewed hermeneutics as a problem of logic.

Many important philosophers contributed to the impact of Enlightenment hermeneutics, including Konrad Dannhauser in his *The Idea of a Good Interpretation* (1630);[4] J. G. Meister's influential *Dissertation on Interpretation* (1698);[5] and perhaps most importantly, Johanne Heinrich Ernesti's, *On the Nature and Constitution of Secular Hermeneutics* (1699).[6] These texts raised many Enlightenment issues about interpretation and contributed to the formation of hermeneutics as a distinct philosophical discipline.

Christian Wolff

One of the most prominent eighteenth-century philosophers was Christian Wolff. Like many of his contemporaries, his major work was on logic. However, within this were several highly influential chapters on interpretation, entitled, 'On Reading Historical and Dogmatic Books', and 'On Interpreting the Holy Scripture'.[7] Wolff dedicated a serious proportion of his philosophical studies to interpretation as he believed it was of pivotal importance. He thought that most of our knowledge was received

[2] See Kant 1996; 1997.
[3] See Gadamer's later use of this question, Gadamer 1981; 1989.
[4] See Dannhauser 1630.
[5] Meister 1698.
[6] See Ernesti 1699.
[7] See Wolff 2003.

through books so that it was crucial that we learned the art of critical judgement about the truth content of these books. Did these books contain adequate representations of the truth; or were they false, incomplete or merely spurious?

For Wolff, works, in what we would refer to as the humanities and arts, came in two distinct categories. First, there were historical texts recording the major events in our past. Secondly, he proposed that other literature was of a 'dogmatic' nature, by which he meant it provided us with an understanding of truths pertaining to normative issues of how we ought to live and about metaphysical and religious issues. According to which of these two categories of written work one was dealing with, the interpreter should adopt different criteria for critical judgement. For example, he believed that, on the one hand, historical writings ought to aim for a completeness of account, truth and sincerity, whilst, on the other hand, dogmatic writings should be judged by the quality of their argument, the author's knowledge of the subject matter and again, the truth content itself.[8]

For Wolff, texts should be written in a clear and logical fashion, developing consistent definitions and a clear and logical writing style. If this was achieved, there would be little difficulty in understanding the meaning of the text. The writing of the text and the interpretation would both proceed smoothly and understanding would be gained if the rules of logic were clearly followed.[9]

For Wolff, authorial intention is key to understanding. By authorial intention he means something objective, not subjective. That is to say, the authorial intention has nothing to do with the author's feelings or personality but rather it relates to the genre he is intending to write in. For instance, does he intend a historical account of a naval battle, or a dogmatic treatise on morals? If the former, then different rules should be adopted to judge the success of his project than if he were judged to be writing the latter.[10]

Wolff not only divided writings into the broad genres of historical and dogmatic, he also subdivided these broad categories into more specific ones. For instance, he broke down historical writing into categories like natural history, church history, political history and so on. Each of these had a distinct set of rules to be followed and the critical judgement of the

[8] Wolff 2003.
[9] Wolff 2003.
[10] Wolff 2003.

text revolved around how well the author lived up to these rules. These rules, as with all writing, were universally applied to each category.[11]

Chladenius

The universalising rules of Enlightenment interpretation were transmitted to Wolff's successor, Chladenius. Johann Martin Chladenius was born in Wittenberg in 1710. He studied and taught there in the fields of philosophy and theology and became professor in 1742. He moved to Erlangen where he was professor until his death in 1759. During his relatively short forty-nine years, he was prolific and wrote in theology, history, philosophy and pedagogy. Among his texts there were those that were central to the formation of Enlightenment hermeneutics and included his *Science of History* (1752) and *New Definitive Philosophy* (1750).[12] However, it was Chladenius's *Introduction to the Correct Interpretation of Reasonable Discourses and Writings* (1742), which was the first systematic work ever published in German on the philosophy of interpretation.[13] As Chladenius himself expresses it:

Very little knowledge of this discipline can be found in the field of philosophy. It consists of a few rules with many exceptions.... These rules, which were not allowed to be regarded as a discipline, were given a place among the theories of reason. The theory of reason deals with matters pertaining to general epistemology and cannot go into the area of history, poetry, and other such literature in depth. For this is the place of interpretation and not a theory of reason. Hermeneutics is a discipline in itself, not in part, and can be assigned its place ...[14]

Chladenius thereby stakes a claim for hermeneutics as a distinct philosophical discipline in its own right. In many senses Chladenius's own hermeneutics embodied the principles outlined in Wolff's work, although he offered a much more in-depth and detailed study. Like Wolff, Chladenius aims to uncover a rational system of rules governing interpretation. He dedicates several passages to making this point. For example he writes:

In this and in other cases where we do not fully understand a text we need an interpretation. The interpretation is different in each instance so that another interpretation must be used for every dark or unproductive passage. The interpretation

---

[11] Wolff 2003.
[12] See Chladenius 1969.
[13] See Chladenius 1969.
[14] *Introduction to the Correct Interpretation of Reasonable Discourses and Writings*, 1742: 177, in Chladenius 1985.

may express itself in an infinite number of ways, but, *just as all repeated human actions proceed according to certain laws, an interpretation is also bound by certain principles* which may be observed in particular cases. It has also been agreed that a discipline is formed if one *explains, proves,* and *correlates* many principles belonging to a type of action. There can be no doubt, then, that *a discipline is created when we interpret according to certain rules.* For this we have the Greek name 'hermeneutic' and in our language we properly call it the art of interpretation.[15]

Like his predecessor, Chladenius also focuses his attention upon historical writings. He believes that interpretation is the practice of the humanities, not philosophy, which follows logic. Interpretation is the 'method' used in understanding poetry, rhetoric, history and the Ancients. Chladenius wanted to discover a *universal theory* of interpretation for these humanities disciplines, including a set of practical rules about how to practice interpretation.

Chladenius begins with a definition of hermeneutics. Paraphrasing Wolff, he calls hermeneutics "the art of attaining the perfect or complete understanding of utterances whether these be made in speech or writing". In his own words, he expresses it thus: 'Unless pretence is used, speeches and written works have one intention – that the reader or listener completely understands what is written or spoken'.[16] He proposed two criteria for the attainment of this perfect understanding. The first was based upon an understanding of the author's intentions. (Recall that the author's intentions in the Enlightenment were those of intended genre and nothing at all to do with personality or other subjective attributes). The second criterion for perfect understanding was that of the rules of reason. Understanding requires following the clarity of the language and the rationality of the argument.

In addition to these typical Enlightenment features of hermeneutics, Chladenius introduced a further and somewhat unique point. He focussed upon what he referred to as the 'point of view' or 'perspective' of the observer. Now this might, in our minds, add a relativistic component to hermeneutics, which would be quite at odds with the broader tenet of Enlightenment's foundationalism. However, we can rest assured that Chladenius intended no such thing. His notion of points of view merely reflects that observers will view historical events from a particular stance, but this in no way interferes with the overall objective nature of

---

[15] *Introduction to the Correct Interpretation of Reasonable Discourses and Writings,* 1742: 176, in Chladenius 1985 (my emphasis).

[16] *Introduction to the Correct Interpretation of Reasonable Discourses and Writings,* 1742: 178, in Chladenius 1985.

the event being observed. Moreover, it does not mean we can introduce a relativistic criteria in judging their work. The standard by which historical writings are judged is still one pertaining to the accuracy of the objective truth content of the events depicted. What Chladenius intends by his notion of 'point of view' of the observer is indebted to Leibniz's *Optics* and also bares many similarities to his *Monadology*.[17] Although different historical writers may be observing the same event, they will be doing so from different angles or perspectives, hence their texts will vary somewhat. Moreover, the interpreter must take this into account when judging their works. This is not to say that the truthfulness of their accounts can not be assessed. As Cladenius expresses it, 'Clearly an interpretation has to be correct'.[18] Universal rules of interpretation are still applied to gain a correct interpretation. Indeed, the particularities of the observers' points of view are subsumed into these universal rules.

Chladenius expresses the Enlightenment thrust for universal rules of interpretation in the following passage. He writes that:

Unrefined forms of misrepresentations were used before even a few principles of interpretation had been acknowledged and presented to the various disciplines. Many men of some reputation presented their thoughts as the true opinion and interpretation of the author, taking no interest in being able to account for these interpretations. . . . An attempt was made to slowly restrict these arbitrary interpretations through *rules*, and from these grew the principles of hermeneutics.[19]

Chladenius could not have predicted the complete overthrow of many of the principles of interpretation that he had lovingly woven according to the Enlightenment's intellectual values. For these later developments in hermeneutics we need to look to the new century in Europe and the advent of Romanticism. This movement is one of the most important in the constitution of modern interpretative debates and practices.

ROMANTICISM

The influence of Enlightenment ideas upon hermeneutics was both carried forwards by later nineteenth-century developments and also challenged. Enlightenment understood as an intellectual movement was, of course, associated with science and logic, with the development of

---

[17] Leibniz 1991.

[18] *Introduction to the Correct Interpretation of Reasonable Discourses and Writings*, 1742: 192, in Chladenius 1985.

[19] *Ibid:* 195, in Chladenius 1985 (my emphasis).

rational systems of understanding in all branches of the sciences and humanities. In contrast, the nineteenth century in Europe is often associated with what some have characterised as the 'counter-Enlightenment' movement.[20] This was a group of intellectuals and artists who found the cultural emphasis upon reason both constraining and repressive of the human spirit. They were motivated instead by the elements in human life beyond reason such as art, the passions, imagination, spontaneity, spiritual meanings, ritual and symbols. Moreover, they believed that the particular was more valuable than the universal and they eulogised this feature in folk cultures, the multiplicity of human language, regional identities and the uniqueness of individuals. Furthermore, whereas the Enlightenment was interested in progress, the counter-Enlightenment was often inspired by nostalgia and the reputation of the thinkers from the past. This counter-Enlightenment movement was dominated by Romanticism, and it is in Romanticism that we find some of the most important historical developments in hermeneutics.

It would be a mistake to think that Romanticism was solely a movement against Enlightenment. There are, in fact, certain strands of influence and continuity. Romantic hermeneuticists, for example, claimed that when they taught in the universities their students were no longer satisfied with merely being told what a text meant, but wanted to know *why* that reading was valid. As we have seen, Kant had questioned the possible conditions for knowledge. Romantic hermeneuticists mirrored this and questioned the possible conditions for the interpretation of a text. In so doing they echoed the Enlightenment's concern with questions of the legitimacy of knowledge and understanding.

A more common thrust of Romanticism was, however, its reaction against Enlightenment. This had myriad consequences for hermeneutic practice and theory. One such development that took place at this time was in the essential spirit of Romanticism. Being inspired by art more than reason, the Romantics looked in general to the non-rational aspects of human lives including emphasis upon cultural phenomenon like ritual and symbol. The Romantics were interested in folklore, language, social practices and possible meanings within history. Consequently, hermeneuticists became interested in the interpretation of these cultural dimensions of our lives. Romantic hermeneutic concerns thus

---

[20] Isaiah Berlin was particularly fond of this expression 'counter-Enlightenment'; see for example, Berlin 1976, 1989. It is a slightly broader conception than the Romantic movement itself.

became very broad indeed. How does one interpret all meaningful discourse, that is, conversation, social practices, historical events, cultural symbols and works of art?

Finally, the Romantics infamously rejected the Enlightenment predisposition towards universalisation. Instead, they embodied a focus upon individuality, personal expression, uniqueness and ultimately, genius. Hermeneutics took this on board and applied it to the interpretation of texts. Texts were not merely an instance of a truth content, albeit founded upon doctrine or reason, texts were representations of *individuality* and its *expression*.

We shall see how all these Romantic developments were heralded in the work of the prominent hermeneuticists of the era, by taking a more detailed look at their work. There were many thinkers around the time of Romanticism, or indeed, connected to the Romantic movement – some intimately, others had works which bore a looser affinity. However, of these great names, those that most significantly undertook the incorporation of the ideals of the movement into hermeneutics were F. D. E. Schleiermacher, Wilhelm von Humboldt, Johann Gustav Droysen and Wilhelm Dilthey.

Schleiermacher

Schleiermacher lived from 1768 to 1834. He was one of the co-founders with Humboldt of the University at Berlin (1808–10) where he taught until 1834. Accepted as the creator of *modern* hermeneutics, he was a German philosopher whose talents spread into many further realms. For example, his was the classic translation of Plato into German. Moreover, he was involved with modern theological and religious studies and an advocate of woman's rights. His membership of the Early Romantic circle spanned from 1796 to 1806, and he was, arguably, their most prominent member. His works consisted mainly in texts from his lectures and are entitled '*Hermeneutics: The Handwritten Manuscripts*'. They are collected and translated from the 1980's into no less than a forty-volume edition.[21]

Schleiermacher's concerns, like those of the other Romantics, encompassed language and history. Moreover, when it came to issues of understanding, against his biblically orthodox or enlightened predecessors, his was not a preoccupation with either doctrinal or rational truth. His was a concern with the Romantic passion for the uniqueness of individual

---

[21] See Schleiermacher 1977.

expression. This represented a seminal shift: understanding moved for the first time away from finding an absolute truth, to centring upon issues of individual creativity.[22]

The key hermeneutic concern of Schleiermacher's was that of how to gain insight into the understanding of other people's expression. That is to say, he was interested in how to understand from reading a book what exactly the author meant by it. He was not worried whether what they said was objectively true or false. For example, if I followed Schleiermacher's approach, I would no longer want to know if it was objectively true whether the world was created in seven days. I would want to know what was *meant* or *intended* by the author in saying this. Likewise, I would not wish to discern if it were objectively true that Christ was 'the light', I would desire to know what Christ himself as a thinking, feeling, unique individual had meant when he said 'I am the light'. Turning to political thought for a further example, I would not be concerned about whether Rousseau's *Social Contract* revealed objective truths about political systems; whether it gives good political advice, or indeed if any of his ideas were practical at all. I would simply want to know what Rousseau *meant* or intended when he wrote it. As a Schleiermacherian hermeneuticist you care about what the author *meant* subjectively.

Schleiermacher, as a hermeneuticist, had to turn these concerns into practice. He had to invent ways in which we could decipher the author's own meaning and intention from a text. The result was a new form of hermeneutic practice. Schleiermacher's new hermeneutics had two key elements. Interpretation consisted in a *grammatical* and a *psychological* element. The grammatical element was a technique that addressed the practice of how to interpret the language and form of the text. The psychological dimension, meanwhile, was concerned with gaining access to the author's intentions and meaning.[23]

First, in order to interpret a text, the grammatical component consisted in a close examination of the language used in the text – the meaning and etymological origin of words. Moreover, the language had to be understood in relation to an *oeuvre*. Furthermore, besides linguistic issues, the work should be placed in its context. This is twofold. First, the context of the literary genre should be borne in mind. Thus, if, on the one hand, we were to interpret a nineteenth-century novel, we should know something of that particular genre of literature. If instead,

---

[22] See Forstman 1977.
[23] Schleiermacher 1977: Introduction and Parts 1 and 2.

on the other hand, we wanted to interpret a political tract from the seventeenth century, we should learn first about its own distinct genre. Second, the historical context must be taken into account. In Schleiermacher's words: 'First canon: A more precise determination of any point in a given text must be decided on the basis of the use of language common to the author and the *original* public'.[24] He continues, 'the era in which an author lives . . . wherever these factors make a difference in a finished text – constitute his sphere'.[25] A further aspect to the grammatical dimension of interpretation is that of assessing the internal structure of the text itself. Herein, the meaning of individual words or phrases should be deciphered through relating the *parts* of the text to the *whole*. For example, at a trivial level, if we want to understand a sentence about the description of a place, we need to read further on in the text when details about where and what that place may be, become apparent. The sentence may refer to a city in modern America, it may be an ancient city, a town in modern Europe, or, if the work is a fantasy novel, it might be an imagined place. Of course, some of this interpretative work we do without being conscious that we are in fact doing it. Schleiermacher expresses it thus: 'a sentence regarded in isolation has a sense [*Sinn*], but not a purport [*Verstand*], for only a complete text has a purport.[26]

Schleiermacher's second dimension of interpretation is known as the *psychological* one. This is where the scholar attempts to understand the author's intention and meaning. It consists first in a knowledge of the author's life. You have to know the biographical details of when and where he was born, his family history, the key events of his life, his occupations, travels, friendships. So, too, any good understanding of an author must include reference to the culture within which the author lived. The place of his birth is consequential, and most importantly, the times in which he lived: it is important to know the historical era of the author's life and the epic events that would have shaped his environment. Moreover, the cultural meanings, including the meaning of language within his own period, are most significant. Schleiermacher declares: 'The ultimate goal of technical interpretation is nothing other than a development of the beginning, that is, to consider the whole of the author's work in terms of its parts and in every part to consider the content as what moved the author and the form of his nature moved by that content'.[27]

[24] Schleiermacher 1977: Xii, 1. Part 1 (my emphasis).
[25] Schleiermacher 1977: XII, 3.
[26] Schleiermacher 1977: XII. 2.
[27] Schleiermacher 1977: Part 2: 2.

The second element of the psychological aspect of interpretation is known as *psychological interpretation*. About this, Schleiermacher writes: 'Just as every act of speaking is related both to the totality of the language and the totality of the speaker's thoughts, so understanding a speech always involves two moments: to understand what is said in the context of the language with its two possibilities, and to understand it as a fact in the thinking of the speaker'.[28] The truly original and Romantic element to Schleiermacher's hermeneutics is the notion of *psychological divination*.[29] As we have said, the Romantics were interested in individuality, and the Romantic hermeneuticists in the meaning of a text as *intended* by the *individual* author. Psychological divination was the way in which Schleiermacher thought that we could gain insight into the meaning intended by another individual.

Psychological divination is a kind of *empathy* or *identification* with the author. If the interpreter wishes to understand the meaning of what the author has said, the best way he or she can do this, according to Schleiermacher, is to empathise with the author. That is, the reader must imagine him or herself to be the author and then try to experience what the author experienced: try to think and feel as he or she thought and felt. Through trying to experience what it felt like to be the author, only then can the reader gain apprehension of what the author meant and intended. In Schleiermacher's own words, in psychological divination the interpreter: 'transforms himself into the other and seeks to comprehend individual immediacy'.[30] For example, if we want to know what Machiavelli's text '*The Prince*' means, we need to find out as much about the author as we can (who Machiavelli was, when he lived, what his life experiences were etc.). Then we need to find out about his personal life, his feelings, his imagination, aims, ambitions and motives. Once we have done all that, we should then attempt to 're-live' this. In so doing we come as close to empathy and identification with Machiavelli as we can. Then, through this psychological divination we gain a glimpse into what Machiavelli might have felt and intended when he wrote his book.

In sum, for Schleiermacher, interpreting a text consists, on the one hand, of grammatical interpretation, knowing the language, genre and body of works of the author. On the other hand, it consists in psychological divination, reconstructing an author's original experiences and intentions and then psychologically identifying or empathising with them. As

---

[28] Schleiermacher 1977: Introduction, II. 5.
[29] Schleiermacher 1977: Introduction, II. 5–6.
[30] Schleiermacher 1959: 109.

he himself wrote: 'These two hermeneutical tasks are completely equal, and it would be incorrect to label grammatical interpretation the "lower" and psychological interpretation the 'higher' task'.[31]

Schleiermacher's views about hermeneutics were influential to many of his Romantic and humanistic contemporaries. Of these figures there were those who developed theories derived from their larger body of works, often looking to linguistics and the philosophy of history for inspiration. Wilhelm von Humboldt is typical of this style of influence and his work stands out from many of his contemporaries for its originality and subtlety.

Wilhelm von Humboldt

Wilhelm von Humboldt was born near Berlin in 1767. His brother, Alexander von Humboldt, became the famous scientist and explorer. Wilhelm himself studied law, classics and philosophy. After an initial career serving in the judiciary he moved to Jena where he studied broadly across the humanities and arts, contributing to aesthetics, literature, philosophy and anthropology among other disciplines. He moved in diplomatic and political circles, and also was close to Schiller and Goethe. In the end, it was to his linguistic studies that he focussed his main attentions, and moreover, within these he gave special attention to issues in hermeneutics. Indeed, his writings on poetry, literature and language all contain seminal hermeneutic ideas. Furthermore, Humboldt, like others of his era, found the study of history for its humanistic content essential within his influential historical writings he also displayed a highly learned concern for hermeneutics and proposed some new and influential ideas.[32]

Looking first at Humboldt's contribution to hermeneutics through his writings on linguistics and the philosophy of language, we see a shared belief with the other dominant Romantics of the time. Like Schleiermacher, A. W. Schelling, F. Schelling and Novalis, Humboldt held the general Romantic linguistic viewpoint. He conceptualised human nature as intrinsically linked to language and to an essential structure of the mind.

In his philosophy of language, he borrowed certain ideas from Schleiermacher. The latter had conceived of human language as both an

---

[31] Schleiermacher 1977: III, 7.
[32] See Mueller-Vollmer 1985: 12–17; 98–99.

(abstract) system and as 'utterances', or speech. To this view, Humboldt added two further ideas. He wrote: 'At this point I shall discuss the *process* of language in its broadest sense. I shall consider it not merely in its relationship to speech and the stock of its word components as its direct *product*'.[33] That is, he introduced the notion of language as a *process* and as a *product*. Furthermore, he maintained that language could be both process and product whether analysed either as a system or as the activity of speech. He therefore laid great emphasis upon both the dynamic act of speaking and the results created by that act. However, against these more universalising strands in his philosophy of language, he added characteristically Romantic elements.

First, language and speech although containing universal features also held elements that were particular. They were particular to society. Indeed, every language was unique and embraced unique ways of understanding the world associated with them. Humboldt proclaims the link between language and nation in his phrase: 'Since the *variation* of languages is based upon their form, and since the latter is most closely associated with the intellectual capacity of *nations*'.[34] Herein, Humboldt clearly contradicts the non-nationalist, universalising principle of Enlightenment.

Secondly, just as language was unique to particular societies, so too was it unique to individuals. Thirdly, these ideas of uniqueness were carried further with the idea that every speech utterance itself could be regarded as an individual creation.

Humboldt's philosophy of language naturally paved the way to hermeneutic issues, as he believed that understanding was principally linked to language. In 'The Nature and Confirmation of Language', in *Introduction to the Kawi language* (posthumously, 1836), Humboldt argued that understanding derived from man's essential linguistic nature:

Language is the formative organ of thought. Intellectual activity... becomes externalised in speech and perceptible to the senses. It and the language, therefore, form a unity and are indivisible from one another. Intellectual activity is inherently tied to the necessity of entering into a combination with the phoneme (*Sprachlaut*).[35]

Understanding in this view, therefore, should not be regarded as an 'external' skill but as an essential characteristic of human beings. Indeed,

33 Humboldt 1971: 'The Nature and Confirmation of Language': 1 (my emphasis).
34 Humboldt 1971: 'The Nature and Confirmation of Language': 1 (my emphasis).
35 Humboldt 1971: 'The Nature and Confirmation of Language': 2.

understanding was inextricably linked to the central feature of human beings' ability for speech itself. Speech, for Humboldt, was a fascinating activity that was in part intellectual and in part social. Therefore, just as speech and language had certain universal and certain particular features, so too did understanding. Forms of understanding varied between different societies and so too did the very ability to understand at all.[36] For Humboldt, understanding is not regarded as a merely specialised activity. It is made possible by human community and is grounded in language. He expresses this when he writes: 'By the same act through which he spins out the thread of language he weaves himself into its tissues. Each tongue draws a circle about the people to whom it belongs, and it is possible to leave this circle only by simultaneously entering that of other people'.[37] Furthermore, the sophistication of understanding is related to linguistic competence as it occurs in both the speaker and the listener.

These ideas of Humboldt's, and echoes of similar ideas in the other Romantics, may seem commonplace to us now. However, they were radical at the time for strongly confronting the Enlightenment view of understanding as somehow acquiring an objective meaning that lies 'out there' as an objective other. Language in this view was regarded as a medium that transported objective meaning from one mind to another. For Humboldt, however, meaning in no sense lies 'out there' as an objective other. Meaning lies only within human mind and within human speech itself: 'Inasmuch as thought in its most typically human relationships is a longing to escape from darkness into light, from limitation into infinity, sound streams from the depths of the breast to the external ambient'.[38]

Humboldt's emphasis upon the uniqueness of language, and relatedly meaning and understanding also paved the way for a theory of the limitations to the possibilities of understanding and interpretation, as well as to its possibilities. For if meaning is unique to cultures, individuals and even particular utterances themselves, then every act of understanding can also be an act of non-understanding as each mind is unique *from* the other. This idea left a problem for future generations of hermeneuticists to grapple with.[39]

Besides his theory of language, the second dimension to Humboldt's work, which was central to his hermeneutics, was his approach to history.

[36] Humboldt 1971: 'The Nature and Confirmation of Language': 1.
[37] Humboldt 1971: 'The Nature and Confirmation of Language': 7.
[38] Humboldt 1971: 'The Nature and Confirmation of Language': 2.
[39] Humboldt 1971: 'The Nature and Confirmation of Language'.

Many of the principle ideas of this approach were expressed in '*On the Task of the Historian*' from an address to the Berlin Academy of Sciences given in 1821.[40] Herein Humboldt presents a theory of understanding as part of a theory of history and historical understanding.

The first point he wants to tackle is the issue of how historical understanding is possible. How are we able to interpret and understand the minds, acts, events and cultural forms handed down to us through history? Indeed, how are we able to understand the past at all?

A historian, Humboldt believes, draws upon events themselves. His comprehension, as Humboldt expresses it, 'presupposes, as a condition of possibility, the existence of an analogue' in the person who is comprehending and in the object to be comprehended. That is, he says, a 'recursive primary correspondence between subject and object'[41] is needed. Whereas in linguistic understanding correspondence is provided by the commonality of language, historical understanding also requires this correspondence. But how does this exist?

It exists, in fact, Humboldt claims because the historian himself is part of history and history is active within him. This forms a 'pre-existing basis of understanding', that is to say, an inner bond between the spirit of history and the historian. This bond is between the historian as subject and the historical object to be understood and can be seen to form a commonality between them.[42]

Humboldt also addresses the mode of interpretation used by the historian. He points out the similarity between the creativity of the poet and artist and that of the historian. Especially, he claims, the historian in the act of understanding engages in an act of creativity parallel to other artists. What the historian sees, Humboldt claims, is only isolated particularities. He never perceives the linking nexus. Indeed, it is he who has to supply the picture of the whole and to unite particularities into an ordered unity. Also, he has to create a meaningful totality out of apparently meaningless particularities, and thereby supply the meaning behind these isolated particularities. In this, there is a parallel with the creativity of the artist: both use their creative imagination to produce a meaningful whole. Because of the need to produce a meaningful whole out of isolated particularities, Humboldt believes the historian must study the form of history. That is to say, not merely the events themselves are

[40] Humboldt 1968: Vol. IV: 48.
[41] Humboldt 1968: Vol. IV: 48.
[42] Humboldt 1968: Vol. IV: 'On the Task of the Historian'.

significant, but the arrangement and structure in which they appear and link together.[43]

In short, Humboldt uses ideas from his own philosophy of language and history and fuses these with notions about the centrality of creativity in human understanding to create a truly Romantic hermeneutics with many parallels to Schleiermacher's own. He was succeeded by other Romantics who developed hermeneutics in differing ways and directions. Most notable of these was Droysen, who contributed towards methodological historical questions.

Johann Gustav Droysen

Johann Gustav Droysen was born in Treptow, Pomerania, in 1808. He attended the University of Berlin and taught there from 1833. He wrote Greek translations and significant historical theses including the *History of Alexander the Great* (1833),[44] and the *History of Hellenism* (1836–43).[45] He became professor at the University of Kiel and there involved himself in politics, but later moved on to Jena before again moving on in 1859, this time to Berlin. He wrote seven volumes on the history of Prussia (1855–84) and was unusual for such a prolific historian in displaying a profound interest in methodological and philosophical problems, that is in 'historiography'. His collected lectures on this topic were known as the *Encyclopaedia and Methodology of History* and was published in 1858 as *Historik*.[46] Droysen actively taught and developed the seminal ideas in his *Historik* over the course of three decades until his death. He died in Berlin in 1884.[47]

Droysen's influences included the familiar figures inspirational to Romanticism, including Hegel, Schleiermacher and Humboldt. His thought offers strong parallels with the 'humanist' thought of Vico, and, as we would expect with all these particular influences, his feelings about issues in the practice of history are strongly in the anti-positivist camp.[48]

To support his 'anti-positivist' stance, Droysen developed the idea of three separate forms of inquiry. First, there was speculative inquiry, with the aim of 'finding out'. This belonged to the disciplines of philosophy

43 Humboldt 1968: Vol. IV: 'On the Task of the Historian'.
44 Droysen 1917.
45 Droysen 1877.
46 Droysen 1977.
47 See Mueller-Vollmer 1985: 118–119.
48 See Mueller-Vollmer 1985: 17–20.

and of theology. Secondly, there existed the physical realm of study wherein the task was to provide explanation. Finally, we had the third category of analysis, which was the historical one, wherein the task was to attain, not 'explanation' but 'understanding'.[49]

Droysen's emphasis upon understanding belies the fact that he believed history could never be practiced in an *objective* fashion. No written or indeed spoken account of history could ever lay claim to have captured events strictly as they occurred in the past. From the fragments of historical data and evidence handed down to us in the present, we could not hope to piece together a complete or objective account of the events that were recorded in those fragments. That is, expressed in contemporary terms, history could never lay claim, nor indeed should aspire, to be a science: scientific methods and goals of 'explanation' should not trespass into the historical field. The historians task, if not a scientific one, was Droysen argues, an interpretative one: the historian should seek to attain some understanding through the evidence available to him.[50]

In his non-scientific or humanistic stance towards history, Droysen parallels Vico. In his *New Science* of 1724 Vico made the point that humans can only ever really understand the world of history because this is made by humans and therefore is in some sense internal to us. This makes history distinct from the natural sciences and following on from this, it is therefore necessary to adopt modes of understanding in history specific to its own nature.[51]

Although Droysen knew that humans had made history and that it was therefore somehow internal to us, he had to figure out exactly what the nature of this human creation was. German philosophy was developing fast in the field of philosophies of history. It was Hegel who at this time was leading the world with an astonishing and epic encapsulation of the entirety of Western history.

We have mentioned the influence of Hegel upon Droysen. This line of trajectory was a complex one for Droysen both borrowed key ideas from Hegel's philosophy of history, indeed these were formative in Droysen's overarching view of history. However he also rejected and strongly opposed major elements of Hegel's work.

Droysen had the great privilege of attending Hegel's lectures on the philosophy of world history, a privilege which we would all envy today.

---

[49] Droysen 1977.
[50] From 'History and the Historical Method' in Droysen 1897.
[51] Vico 1982.

From Hegel's notions Droysen was inspired by the idea of a *Geist* – a spirit that was more than subjective and had semi-objective features – to account for social, political and cultural objects. Droysen also was influenced by the idea of an ethical force to history, and adopted this notion in his own work under the term 'ethical powers'.[52]

In spite of being impressed by the notion of a semi-objective spirit to our history embodied in our social, cultural and political institutions, and in spite of accepting the notion that world history had an ethical dimension, Droysen rejected Hegel's key notion of teleology. Droysen did not accept that history was teleological and that as the forces of history unravelled, we would, at history's end point, all attain rational 'consciousness' and inhabit a fully rational society. That is, he rejected both the idea that history would attain completion, and that through the development of world history, complete understanding would emerge. Droysen, in fact, thought that social communities during the course of history were more likely to only partially achieve ethical ideas. Moreover, he was more influenced by Humboldt and Schleiermacher's hermeneutic notion of understanding.[53]

In fact, for his hermeneutic way of conceptualising history, Droysen borrowed from the linguistic modes of understanding indebted to Romantic linguistic theory. These made a strong parallel between historical and linguistic understanding, and even went so far as to proclaim that: 'Our historical understanding is quite the same as when we understand someone who is speaking to us'.[54]

Historical understanding for Droysen proceeds first, through assessing the truth of the historian's data and evidence. Secondly, it proceeds through the interpretation of this data. This hermeneutic task has several dimensions which Droysen terms 'pragmatic', 'conditional', 'ethical' and 'psychological'. We have seen that the ethical stance implied in historical understanding comes from Hegel, whereas aspects of the other dimensions arise from earlier Romantic theories of interpretation. However, the psychological part of hermeneutics is new and is Droysen's own original contribution.[55]

This original element builds upon the Romantic idea that history can be understood in the same way that speech is understood. Consider the

[52] Droysen 1897.
[53] Droysen 1897.
[54] Droysen 1977: 35.
[55] Droysen 1897.

points made in Humboldt's *Introduction to the Kawi Language*, especially his notion of meaning residing in 'utterance'. Herein Humboldt had held the idea that because meaning resided in speech (or utterance) understanding proceeded by uncovering the meaning of each utterance.[56] Although Droysen accepted this, he looked well beyond the actual rational or semantic meaning that Humboldt had focussed upon, and gave his attention to the *expressive* functions of language.

Droysen hereby focussed upon hermeneutics proceeding through an additional key element which we can see to be psychological, emotional and spiritual. An utterance contained an expression of something internal to any individual's state of mind. It disclosed an attitude or state of mind of the originator and Droysen believed that we had to understand this in order to understand the utterance itself. He wrote:

The possibility of this understanding arises from the kinship of our nature with that of the utterances lying before us as historical material. A further condition of this possibility is the fact that man's nature, at once sensuous and spiritual, speaks forth every one of its inner processes in some form apprehensible to the senses, mirrors these inner processes, indeed, in every utterance.[57]

Droysen then combines the emphasis upon accessing the psychological state of mind of a speaker with the view that historical interpretation proceeds through a similar means as its linguistic counterpart. However, although the 'method' of moving between the parts and imagined whole of historical data in order to try to accomplish interpretation was highly significant to other romantic views, Droysen having rejected Hegel's notion of a teleological unity, also criticises the idea that we can assume an underlying unity to history.

These features and problems in Droysen's attempt at formulating a specifically historical hermeneutics were passed on to his colleagues and successors. Foremost of these, and highly influential throughout both the Continent and British philosophical and historical circles today, was Dilthey.[58]

Dilthey

Dilthey was a figure who was of enormous import to Romantic hermeneutics. He lived from 1833 to 1911. Like Schleiermacher, Dilthey was a

---

[56] Humboldt 1971.
[57] Droysen 1897: #9.
[58] See for instance Skinner 1988.

German philosopher, and was also a historian. He studied theology, history and philosophy at Heidelberg and in Berlin. A hugely influential figure, he became Chair in the University of Berlin after Hegel in 1882. Dilthey was influenced by Schleiermacher and this inspiration is clear in many of his works. Among Dilthey's key works are *Introduction to the Human Sciences*, published in 1883;[59] *Ideas*, published in 1894;[60] and *Formation of the Historical World in the Human Sciences*, published in 1910.[61]

Dilthey's hermeneutics builds upon many of Schleiermacher's key concerns. He also, of course, made his own contribution. Throughout his life Dilthey was preoccupied with the need to study ideas in relation to their social context. That is, he wanted to place theories in their social and cultural context rather than studying them as if they existed outside of human lives and time. Great thinkers of the era, and before, largely under the influence of Enlightenment and the impact of the success of the natural sciences, had studied ideas as if they were universal, that is to say, true for all languages, cultures, historical periods, and peoples. In contrast, for Dilthey, a key concern was to make a distinction between laws that were universal and applied to the natural world, and meanings that were historically generated and applied to the social world.

It was in Dilthey's later work that he developed his concerns in the direction of hermeneutics. In this realm he wanted to make a distinction between *explanation* which was the focus of the natural sciences and *understanding* (*Verstehen*) which he believed was the project for the human 'sciences'.[62] The goal of understanding in the human sciences should be pursued, thought Dilthey, in line with hermeneutics in general, through the process of interpretation. 'Understanding and interpretation is the method used throughout the human sciences. It unites all their functions and contains all their truths. At each instance understanding opens up a world'.[63] Like Schleiermacher, Dilthey was a Romantic hermeneuticist. He took his concern with the distinction between the natural and human sciences – and with the historical nature of the latter – to hermeneutics, and applied the approach of Romantic hermeneutics to history.[64]

First, like Schleiermacher, Dilthey began interpretation with 'grammatical' textual analysis. One had to make a linguistic study of texts, be

[59] Dilthey 1989.
[60] In Dilthey 1914–1990 and 1985–.
[61] Dilthey 2002. See also Owensby 1994: 1–21.
[62] See Owensby 1994: 51–79.
[63] The Understanding of other Persons and Their Life Expressions' in Dilthey Vol. 7, 1926.
[64] See Owensby for an excellent commentary.

sensitive to the relation of the meaning of the parts of the text to the meaning of the whole. One had to know the genre of the text and situate it within a body of works.[65]

Secondly, as with Schleiermacher, the process of psychological divination was crucial to interpretation. One had to empathise with an author to know the meaning of their text. One had to attempt to re-live their experiences, motivations and feelings and so on.[66]

The new approach inaugurated by Dilthey was to apply this Romantic hermeneutic approach to historical understanding. Historical understanding was, after all key to any human science. First, the nature of human experience was historical in contrast to the natural sciences where the laws of nature traversed all time. Secondly, the human sciences could not therefore merely transpose the methods of 'explanation' from the natural sciences onto their subject matter. The human sciences needed to pursue the goal of understanding. And this could be achieved through the interpretation of human history.[67]

Hermeneutics, had however, been mainly concerned with the interpretation of the meaning of *texts*. It had under Schleiermacher been shown to be more broadly relevant to all aspects of human culture, but no-one had attempted to generate historical understanding solely in this ambitious way. How could one use approaches of interpretation developed mainly upon texts and apply them to the whole of human history? More specifically, how could one apply psychological divination to human history?

Dilthey approached this in the following way. He conceived of the total socio-historical world as a text. Once the whole of human history is portrayed as a text, then, like the understanding of any grand narrative, we can use the Romantic hermeneutic method. The whole of human history, is after all just like a novel or story, and we and all our predecessors and neighbours are the characters in the story. The events that sweep through and shape our lives are the events of the narrative. So now we can see what interpreting our story of human history consists in.[68]

First, Dilthey takes the 'grammatical' approach. Herein we need knowledge of the language of past peoples and events. We need to be sensitive to the structure of the narrative and interpret the meaning of the parts

[65] Dilthey Vol. 7, 1926.
[66] Dilthey Vol. 7, 1926.
[67] Dilthey Vol. 7, 1926.
[68] 'Awareness, Reality: Time' from 'Draft for a Critique of Historical Reason' in Dilthey Vol. 7, 1926.

of the narrative in relation to the whole. Thus we interpret the particular in terms of the whole. Whereas in a text we would interpret the meaning of paragraphs in terms of the larger overall text, here we interpret the meaning of small events in terms of larger ones, for example, the meaning of a battle in terms of the narrative of larger wars and conflicts.

Secondly, as with textual interpretation more generally, if we require to understand the meaning of the actions of people from the past, we need to learn the details of their lives, that is to gain biographical knowledge and we also need to understand their social and cultural context.

Furthermore, a form of psychological divination, or empathy should occur. Herein, we need to imagine ourselves to be the historical character. For example, we need to 'become' Napoleon in order to feel what he felt; we need to become Napoleon in order to intend as he intended. For another more down-to-earth example, imagine yourself to be a peasant during the English Civil War. How did you think, feel, experience life? When this psychological divination has been accomplished, you would, according to Dilthey, have gained some understanding of your historical world.[69]

Dilthey's fusion of humanism with an anti-Enlightenment focus upon the imagination, typifies the thrust of Romantic hermeneutics as a whole. This tradition in hermeneutics can be seen to be a reaction against the Enlightenment's encroaching emphasis upon science and the search for universal laws. Moreover, it can be perceived to consist, on the one hand, in a humanistic element and, on the other hand, in its own distinctively modern contribution. As we can recall, the Ancient Greek, Roman and early Christian schools of interpretation occurred in either the Antiochian mode of literal, that is factual, often historical readings of texts, or in the Alexandrian tradition of allegorical, spiritually meaningful accounts. The Romantics followed aspects of both of these early humanistic traditions. On the one hand, they were all deeply committed to historical reading and understanding. Moreover, on the other, they practiced highly allegorical modes of interpretation, often indebted to developments in semiotics and philosophy of language going beyond the early Church fathers like Augustine. However, the Romantics bridged the classical divide between historical literalism and spiritual allegory by offering readings that were indebted to an allegorical approach to history.

In contrast to Enlightenment, the Romantics were not only more intimately connected to earlier humanism, they were also clearly indebted

---

[69] 'Empathy, Re-creating and Re-living' in Dilthey Vol. 7, 1926.

to broader intellectual developments from the blossoming philosophical traditions of their times. They were influenced by accounts of culture and society by minds like Herder and Vico, philosophies of history by thinkers like Hegel, and their own developments in historical and linguistic understanding. They further developed a distinctively modern individualist approach to interpretation which pointed to the importance of authorship, genius and creativity.

With the increasing advances and sophistication of German philosophy throughout the nineteenth century, new ideas were developing which were bound to have a significant impact upon post-Romantic hermeneutics. Chief amongst these were the phenomenology and existentialism advanced by thinkers of the early twentieth century like Husserl, Heidegger and the theological schools that followed them. It is to these that we shall now turn.

4

German Philosophical Hermeneutics

Phenomenology and Existentialism

PHENOMENOLOGY AND EXISTENTIALISM

In this section of our study we turn our attention to two eminent and quite complex strands of philosophy, both of which exerted an enormous grip upon hermeneutics. With the advent of phenomenology and existentialism we witness a wholly new shift in interpretation from issues about its practice to deeper underlying concerns about the nature of understanding itself. Although both Enlightenment and Romantic hermeneuticists discussed how we come to comprehend in the first place, and both provided analyses of the linguistic and historical components to this, neither tradition perhaps went quite as far as these two new 'schools' of philosophy. Moreover, Heidegger, actually turned his concern almost entirely towards a philosophy of interpretation and understanding whereas issues of exegetical practice all but faded from sight.

Husserl

Edmund Husserl, born in 1859, is the German philosopher credited as the founding father of phenomenology. He was born in what is now part of the Czech republic, and studied science and philosophy at Leipzig, mathematics and philosophy at Berlin and continued with philosophy and indeed psychology at Vienna and Halle. He taught at many of Germany's major universities until a few years before his death in 1938. Being of Jewish descent, his many papers were in danger from the Nazi regime, and were saved posthumously and now exist in a special archive in Louvain.

Of his many philosophical works, the most ground breaking was his early thesis *Logical Investigations*, 1900–01, in which he delivers his phenomenology.[1] The impact this work has had upon hermeneutics is as far reaching as it is varied. *The Crisis of the European Science* and *The Task of Phenomenology*,[2] his last work is also the bedrock of many strands of phenomenology and has hermeneutic components. It should be noted, however, that Husserl was not himself a 'hermeneuticist' and never actually used the term in his philosophy. However, he is important in two senses, first, because *Logical Investigations* brings the problems of phenomenology to bear upon hermeneutics.[3] In fact, *Logical Investigations* can even be read as the establishment of the foundations and possibility of hermeneutics.[4] Secondly, key points about Husserl's phenomenology should be noted for their impact upon the later Heidegger. It was Heidegger after all who carved from phenomenology and existentialism some of the most radical changes to Romantic hermeneutics, and who, moreover, paved the way for some of the most path-breaking developments in the twentieth-century tradition.

Let us first examine a few features of phenomenology and familiarise ourselves with some of its terms. In fact, as a term, 'phenomenology' is peculiarity vague and frustrating, with many authors apparently taking for granted an assumed coherent meaning. Phenomenonology, however, is rarely defined in any precise way. In general it denotes a dispersed strand of philosophical perspectives rather than a specific school and so as a term is not, in fact, very helpful to us. Almost all strands and varieties of phenomenology do, however, originate from Husserl's works and thus bare certain historical and conceptual affinities.

In Husserl's work, what exactly is meant by the term 'phenomenology' is itself open to debate. We can say that it refers to an approach to philosophy demarcated by a few key features. First of these is an interest in understanding essences. How are scientists able to comprehend essences in the objective world? Science can in fact, according to Husserl, only pursue the study of essences in the objective world because it is itself founded upon meaning existent within the 'life world'. Husserl's phenomenology then goes on to analyse the notion of meaning within the 'life world', rather than focus upon the properties of objective things,

[1] Husserl 2001.
[2] Husserl 1970; also in Husserl 1950–, 1954.
[3] Husserl 2001
[4] See Kockelmans 1967.

which he believes is the duty of science. Husserl therefore demarcates philosophy from science and moreover places philosophy in a foundational role with respect to science.[5]

Second, Husserl defends a form of intuitionism. He proposed that we human beings have a kind of pre-conceptual apprehending of phenomena. Indeed, it is from this intuitive knowledge of essences that the objective sciences are able to develop.[6]

Third, our intuitive apprehending of essences occurs within particular life worlds. These life worlds are composed of historical communities of which there are multiple instances ranging, for example, from Mexican Indians to Welsh farmers. However, Husserl also goes on to investigate the possibility of a common underlying 'life world' which would facilitate the notion of certain universal truths, and indeed allow for the possibility of the objective sciences themselves.[7]

Finally, intentionalism is a key concept for Husserl. This is because he denotes intentionality as a key feature of conscious phenomena, in contrast to acts and events of a non-conscious type which are often unintentional. The idea of intention Husserl believes plays a crucial role in relation to meaning and he explores how intention relates to both verbal and non-verbal acts and moreover, how this forms the cornerstone of meaning. In this vein, he distinguishes between an act, the intention of the act and other forms of content within the act besides intentionality. Thus he is keen to demarcate the specific intention of an act, including verbal acts, from the whole context surrounding that verbal act, for instance from the unintended dimensions, including the reaction of the audience and so on.[8]

Having witnessed a general overview of Husserl's phenomenology we can begin to see how his work might spearhead changes in hermeneutics. First, in demarcating philosophy from science, Husserl removes from philosophy a commitment to scientific epistemologies and methods thereby leaving room open for other non-scientific approaches. That is, he opens the door for a cardinal role for hermeneutics.

Second, Husserl developed the notion of communities with intuitive apprehending of meaning. Thereby his philosophy is concerned with

[5] See Kockelmans 1967.
[6] Husserl 1970.
[7] Husserl 1970.
[8] Husserl 1970.

shared meanings and understanding and so overlaps with fundamental hermeneutic concerns.

Third, in developing the concept of multiple life worlds, Husserl brings in the notion of communities of meaning and therefore the need, and indeed perhaps, too, the very possibility of interpreting these diverse meanings.[9]

Finally, the notion of intentionalism is taken up by later hermeneuticists like the American, E. D. Hirsch to develop one brand of a specifically phenomenological hermeneutics.[10]

In spite of his rigorous assertions for the non-scientific nature of philosophy, Husserl's Phenomenonology is itself, if not an objective science, certainly he claims, a rigorous one. Indeed, his project is closer to science than that of the many who followed him, including his greatest 'student' Heidegger. In fact, the phenomenology of Husserl has led to disparate developments, many of them closely connected to science rather than to works within non-scientific strands of philosophy like the interpretative ones. There are notable exceptions to this, for instance, in the work of Roman Ingarden,[11] Hirsch and of course, Heidegger.[12]

Of the various strands of Husserl's phenomenology, Hirsch uses the concept of intentionality to construct a notion of verbal meaning. Unlike Schleiermacher, who identified meaning with the state of mind of the author at the moment of creating a text, Hirsch equates meaning principally with an author's conscious intention as conveyed through the text itself. Thus, in contrast to Schleiermacher, we don't need to 'psychologically' re-experience what an author experienced at the moment of writing. What we need to do is understand what the author intended in and through his use of language. For Hirsch, this means primarily what the author consciously and deliberately intended in the text. Hirsch writes that meaning is: 'Whatever someone has willed to convey by a particular sequence of linguistic signs and which can be conveyed by means of those linguistic signs'.[13]

Hirsch's definition also, however, allows the meaning of a text to include reference to the objects of the author's meaning, rather than simply the author's mental state itself. That is to say, the meaning of a

---

[9] Husserl 1950–.
[10] See Hirsch 1967 and 1976.
[11] See Ingarden 1975.
[12] See Mueller-Vollmer 1985: 187–193 and Warnke 1987 for more on these developments.
[13] Hirsch 1976. See also Warnke 1987: 44, n3.

text is a combination of the author's conscious intention, the linguistic conventions he adopts and the nature of the object to which he refers. Hirsch explains 'an author's meaning can include more than that of which the author is explicitly aware because the author's intended meaning specifies a certain type of thing as opposed to a particular mental content'.[14]

Husserl's phenomenology, mediated through Hirsch, gave weight to the tradition of hermeneutics. However, for the overriding mainstay of Husserl's work upon the interpretative tradition, we had to await developments inaugurated by Martin Heidegger. It was really Heidegger for whom the thrust of phenomenology was directed in a more specifically hermeneutic direction.

Heidegger

Many of the biographical details of Heidegger's life are opaque, no doubt because of his shameful involvement with the Nazis. We do know, however, that he was born in 1889 in Messkirch in the Black Forest and attended a Jesuit school and indeed first trained to be a Jesuit but then turned to mathematics and philosophy. He worked with Husserl at Freiburg University where he was later to become professor when Husserl vacated the Chair. During the time of the Third Reich (1933–34) Heidegger assumed the rectorship of the University and joined the Nazi party. He taught from 1944 until 1951 when he was suspended by the French military government. However, he continued teaching until 1967, his key work being the seminal *Being and Time* (1927).[15] Moreover, there is under completion a fifty-six-volume edition of his writings.[16]

Heidegger's influences besides Husserl were Kierkegaard and the 'life philosophy' – *Lebensphilosophie* – identified with Nietzsche, Bergson and Dilthey. It is Husserl's influence that we are principally interested in examining here. Heidegger's work occupies an interesting place in hermeneutic history because, for the first time, it places hermeneutics as the principle concern of the philosopher. In fact, Heidegger defines the philosopher's task as an interpretative one and stated boldly that hermeneutics played a foundational role in his own philosophy. This is in stark contrast to the mainstream Western philosophic tradition which

---

[14] Hirsch 1976. See also Warnke 1987: 44.
[15] Heidegger 1962.
[16] See Murray 1978 and Mueller-Vollmer 1985: 214–240.

from Ancient times through to the early twentieth century had made metaphysics, epistemology, logic and, of course, ethics its bedrock.[17]

Heidegger intended *Being and Time* to consist in two parts. The first was an analysis of human existence whilst the second was to be the major evaluation of 'Being' itself. Famously, the second part was never completed, but the first part has proved a seismic contribution to hermeneutic philosophy in its own right.[18]

Heidegger's project is a recasting of the Ancient Greek question of ontology, that is, the nature of the existence of things. In Heidegger's language – he wanted to answer the 'question of being'. However, he wanted to avoid getting trapped into an inquiry into the metaphysics of substance as he believed the Ancients had done. This metaphysical focus, he opined, overlooked the issue of why 'substance' should reveal itself or indeed matter at all. Heidegger wanted to know why things should appear as objects to be known by us in the first place. To answer this question, he turned his attention to the entity that tries to understand, and indeed, has some prior understanding of things in the first place. This entity is human existence, or *Dasein*.[19]

From Husserl's phenomenology we can see, therefore, that Heidegger has adopted the notion that we humans have some kind of intuitive apprehending of the world. In Heidegger's terminology this is expressed as a certain kind of existential understanding that we have as part of our being-in-the-world. Dasein possesses as part of its Being a pre-ontological understanding of self and world.

Heidegger's claim is that Dasein's preontological understanding of everyday things and practices opens up a clearing in which entities can show up, or reveal themselves. The task of the philosopher is to bring to ontological understanding that pre-ontological understanding that Dasein already possesses as part of its being. However, in contrast to other metaphysicians, Heidegger does not wish to take an abstracted 'spectator' attitude towards human existence, *Dasein*, but rather to encounter and describe it in its everydayness. This is our everyday state of existence when we are caught up in our practical affairs, our relations with others and the world, various tasks, our moods and capacity for individuality.[20]

[17] See Murray 1978 and Mueller-Vollmer 1985: 214–240.
[18] Heidegger 1962.
[19] Heidegger 1962.
[20] Heidegger 1962.

What Heidegger discovers through watching us in our everydayness is that understanding is possible because we are understanding entities in the first place. In fact, Heidegger goes so far as to say that we as self-interpreting beings are just what we make of ourselves in everyday life. That is to say, not only are we able to understand – not only is understanding a crucial feature of our existence – but it *is* our existence. To express this cardinal point of Heidegger's philosophy in other words, he is convinced that we are not some priorly given 'essence' that can perform the act of understanding, but we are in fact by our very existence understanding beings. Indeed, our being is, in fact, no more than this very understanding.[21]

Heidegger's gigantic leap in hermeneutics places interpretation as the mainstay of human existence. Further, his philosophy is known as an ontological hermeneutics because it provides an account of how understanding is possible. We humans can understand because our existence is as understanding beings.

In his later work, *An Introduction to Metaphysics* (1935),[22] Heidegger focuses upon language as allowing all understanding. In point of fact, he makes the claim that because language is essential for understanding, human beings do not speak, but rather, language speaks us.[23]

Heidegger's ideas lent majestic weight to later forms of thought, including post-structuralism and postmodernist ideas, most notably in the instance of Derrida. He has also had impact on broader tenets of work like Emil Staiger's *Fundamental Principles of Poetics* (1946),[24] and Ludwig Binswanger's *Basic Forms of Analysis of Human Existence* (1942).[25] Throughout French, Anglo-American and Swiss thought, besides German philosophy itself, Heidegger's ideas are pervasive. Indeed they have made great inroads into theology in the form of the work of the so-called 'New Hermeneutics' of thinkers like Rudolf Bultmann. Because Rudolf Bultmann's work was so heavily indebted to Heidegger, we include him here in our phenomenology/existentialist section, although this is not to detract from his own specifically theological concerns.[26]

---

[21] Heidegger 1962.
[22] Heidegger 2000.
[23] This, of course, is a highly influential idea in later post-structuralist theories.
[24] Staiger 1991.
[25] Binswanger 1963.
[26] See Mueller-Vollmer 1985.

Bultmann

Rudolf Bultmann was born the son of a Lutheran pastor in North Germany in 1884. He studied at Tubingen and then taught at Marburg, Breslau and Giessen, only to return to Marburg as professor until his death in 1976. Bultmann was one of the most important theologians of the twentieth century. His works which gained him this reputation included *Jesus and the Word* (1926),[27] *Theology of the New Testament* (1941; 1951),[28] *History and Eschatology* (1957),[29] and *Jesus Christ and Mythology* (1958).[30] Bultmann was both a great New Testament scholar and also his writings encompass many pivotal topics in theology including hermeneutics.[31]

For his hermeneutics, Bultmann drew upon two sources of inspiration; first, some of the older Protestant hermeneutic traditions discussed earlier in our book; and, secondly, he was highly influenced by Heidegger's *Being and Time*. He was inspired by *Being and Time* in two ways. First, he was influenced by Heidegger's hermeneutic approach to philosophy. Secondly, he adopted many central features of Heidegger's concept of understanding. In spite of his highly accredited biblical scholarship, it is to these latter influences that much of Bultmann's fame is due. Indeed, he is known first and foremost for attempting to go beyond the confines of 'objective scholarship' and actually perform the act of *interpreting* the scriptures to modern man.

Bultmann proposed the notion that the Scriptures contained an existential content, expressed in mythological form and he, therefore, became preoccupied with interpreting through the 'veils' of this myth, the existential message. Indeed, Bultmann perceived the task of the contemporary theologian as one of demythologising Scripture. The contemporary theologian had to interpret behind the mythology of the Bible in order to understand the existential core. Paradoxically, this approach was in many senses contrary to what its name actually seems to imply. For in demythologising, Bultmann does not seek to strip scripture of deeper imaginative elements and reduce the Bible to superficial, literal and readily believable ideas. Rather, he seeks to preserve the most imaginative

[27] Bultmann 1958.
[28] Bultmann 1983.
[29] Bultmann 1957.
[30] Bultmann 1958.
[31] See Bleicher 1980: Chapter 4, Section B for a good account.

meanings from within the text. As Palmer expresses is, Bultmann wants to preserve language 'as the medium through which God confronts man with the possibility of a radically new . . . self-understanding'.[32]

There is much controversy about exactly how much Bultmann was influenced by Heidegger, with some claiming that this relationship has been exaggerated. Nevertheless, it would be hard not to see the striking impact of Heidegger's ideas upon some of the foundational assumptions that underlie Bultmann's project of interpretation. First, like Heidegger, Bultmann makes a crucial distinction between language as information, mere fact gathering, and language filled with 'personal import' and the 'power to command obedience'.[33] Second, Bultmann believes that God confronts man as Word, that is to say, in language. Herein we see the impact of Heidegger's emphasis on the linguistic character of Being. Third, the Word, Bultmann believes, speaks as existential self-understanding. The New Testament, he claims is written in order to bring about a new (authentic) self-understanding in modern man. The emphasis on authentic, self-understanding smacks of Heidegger. Moreover, some, for instance Palmer, view the New Testament Word as similar to the call of conscience that Heidegger describes in *Being and Time*.[34]

Besides his influential approach of demythologising, Bultmann also developed a theory of interpretation of the scriptures. This was based on many phenomenological and existential ideas and was taken up by Bultmann's followers, Fuchs and Ebeling. Together their ideas became known as the 'New Hermeneutic'. Note that a difference emerges here from Heidegger: for the 'New Hermeneuticists', hermeneutic theory is a *guide* to exegesis rather than representing understanding itself.[35]

Bultmann's new theory of interpretation was a significant contribution to hermeneutic thought and took a strong stance towards its central problems. It addressed in detail many practical complexities of reading. At the outset, Bultmann proposed his theory of hermeneutics to be centred around biblical interpretation. This entailed the following. First, the Bible was to be interpreted as a text, in the same manner as legal, historical or literary texts. Biblical interpretation should be subject to the same rules.[36]

---

[32] See Palmer 1969: 49.
[33] Bultmann 1985 (note, in view of the historical era of these writings, the frightening resonance of these ideas of Heidegger's with his later political sympathies).
[34] Bultmann 1985. See also Palmer 1969.
[35] See Robinson 1964 for a good account.
[36] 'The Problem of Hermeneutics' in Bultmann 1955: 252–253.

Secondly, the Bible was to be perceived as predominantly a historical work. The issue therefore arose of not merely understanding the Bible as a text, but crucially, what was meant specifically by historical understanding. This was the central issue at the heart of Bultmann's Gifford lectures of 1955, entitled: What is the character of historical knowledge?'[37]

First and foremost, Bultmann was against the notion of objective meaning in history, be it the shallow literalism of a scientific approach or the semi-objective Hegelian and Marxist teleological view. Historical knowledge could not proceed through the means dictated by these schools of thought. Bultmann was more in sympathy with the interpretative tradition of hermeneutics. Herein, he differed principally from Enlightenment views about objectivity in interpretation itself. Bultmann did not believe that we should look for an objective hermeneutics of history any more than an objective science of it. In fact, against the *tabla rasa* viewpoint of science, wherein we aim to view the past free from any subjective standpoint, Bultmann introduces fairly 'subjective' factors into the interpreters' vantage point.

First, Bultmann believes that the facts from the past only turn into history 'when they become significant for a subject which itself stands in history and is involved in it'.[38] That is, Bultmann claims that the standpoint of the present directs a 'preliminary understanding' of the subject matter of history.[39]

Secondly, not only does the author or historian's temporal position direct the beginning of understanding, but so too does his particular interest in the past. The idea that the author's interests can and do direct interpretation provides a strong distinction from many earlier hermeneuticists – especially those of the Enlightenment. Without these preliminary standpoints of subjectivity of the historical viewpoint of the author and his or her direction of interest, no interpretation would be possible at all, according to Bultmann.

He is strongly criticised by many historians and indeed hermeneuticists from a more scientific disposition. Chief among these perhaps is Emelio Betti, who accuses Bultmann of being completely non-objective about history. Bultmann, Betti believes, has thrown the doors of interpretation wide open to the corruption of subjectivity not only by allowing the

[37] Bultmann 1957.
[38] Bultmann 1955: 252–253.
[39] Bultmann 1955: 252–253.

present to prejudice the past but also by accepting the author's interest as a valid standpoint for the interpretative act.[40]

We have seen how Romantic and existential contributions to hermeneutics are strongly related to theories of language and history. Some interpretative theorists draw predominantly upon philology or linguistics whereas others veer on the side of a historical focus. Most, however, combine the two. Bultmann's approach to biblical interpretation was heavily influenced and shaped by issues in historical interpretation. Whilst his underlying existential demythologising approach, inherited from Heidegger, was appropriated by the new hermeneuticists Ebeling and Fuchs, it was, however, coupled to a shift from a predominant concern with understanding history to the philosophy of language.[41]

The contemporary theologians Gerhard Ebeling and Ernst Fuchs place hermeneutical issues at the centre of their thought. They inherit a concern with a modern rendering of the New testament and oppose a literal reading of scripture. They focus upon the meaning of the Bible rather than its factual content. They, too, are concerned with existential self understanding in the Bible but shift to analyse issues of language.

Ebeling writes: 'Existence is existence through word and in word ... existentialist interpretation would mean interpretation of the text with regard to word event'.[42] Due to this emphasis upon the linguistic nature of theology, Ebeling and Fuchs have been termed 'word-event theologians'. Their shift toward a linguistic style of hermeneutics entails a focus upon language as medium; a vessel of (religious) meaning. Moreover, they focus less on history conceived of as the past and more upon history as the expression of the becoming of the present.

This movement from a historical hermeneutics to a linguistic one is compensated for by the most influential later hermeneuticist who once again appropriates a shift back towards a predominantly historical preoccupation. The influence of Heidegger does however continue, albeit in a more specifically hermeneutic rather than theological vein.

[40] Betti: 1962.
[41] Ebeling: 1959; Fuchs 1959–65.
[42] Ebeling 1959: *Word of God.*

5

Continental Philosophical Hermeneutics Post War

In the final section of our portrayal of hermeneutics we shift our scrutiny to a thorough inspection of the trends of the latter half of the twentieth century. Many of the thinkers addressed here are prominent across all the multifarious disciplines that comprise the contemporary humanities. They exert an almost palpable influence upon philosophical, epistemological and exegetical debates throughout the social and political sciences, literature, history, law and theology. Moreover, their works are influential through being seized upon (rather inappropriately) as methodologies, no more so than in the social sciences. However, we shall continue to see the distinctively humanistic influences of an almost overriding concern with matters of language, tradition and history which distinguish contemporary hermeneutics from the sciences and methodological approaches of the social sciences.

The most influential figure of this twentieth-century interpretative theory is Hans-Georg Gadamer. His ideas inform a score of other thinkers and schools, most notable of whom are Ricoeur and Derrida from structuralist and deconstructionist schools. Furthermore, Gadamer provokes fairly impassioned debate especially from critical theorists like Habermas. We begin with a thorough-going account of Gadamer's own ideas before moving on to assess his successors.

HANS-GEORG GADAMER

Gadamer, in contrast to the other figures discussed, intersects with our own life and times. He was born in 1900 and lived to see the beginning of the twenty-first century. He was a student of Heidegger and he worked in the universities of Marburg and Heidelberg. He officially retired in 1968 but remained very active until his recent death. His principal text, *Truth and Method*, was published in 1960 and translated into English in 1975.[1]

Gadamer was a *conservative* hermeneuticist. His main influences were the discipline of aesthetics, the work of Kant, the tradition of modern (Romantic) hermeneutics and finally, his teacher Heidegger from whom he borrowed both existentialist and conservative arguments. Our purpose in elucidating Gadamer's theory leads us to look mainly at the development of his theory out of the earlier hermeneutic traditions so far discussed.[2]

Kant had defined an epoch with his question, 'What are the conditions for possible *knowledge?*' Gadamer echoed Kant for the hermeneuticists and asked, 'What are the conditions for possible *understanding?*'[3] In his answer to this question, he was first influenced by the Romantics. Understanding occurred through interpretation. Interpretation consisted in two elements, namely, a grammatical and a psychological one. In his description of the latter psychological element he was highly indebted to Dilthey, for he incorporated within it the very historical nuance.[4] Moreover, he was also influenced by critics of Romantic hermeneutics. However, before we address these, let us examine in more detail the overall theoretical apparatus which Gadamer constructed in order to answer his Kantian-inspired question.

To address the issue of the conditions for understanding, Gadamer devised his own hermeneutics. A conceptual account of this can be organised around three points. First, issues of interpretation require a theoretical stance about the nature of the *text* to be understood. Secondly, Gadamer theorised about the nature of the *reader* or *interpreter*. Thirdly he developed a thesis concerning the overall *process* of understanding itself.

The Text

Let us begin our elaboration with a discussion of the first point, namely, Gadamer's notion of the nature of *the text* to be understood. For Gadamer there are key features that we must assume for any text. First, it is an

---

[1] Gadamer 1989.
[2] See the excellent account offered by Warnke 1987.
[3] Gadamer 1989: 42–89.
[4] Gadamer 1989: 28–242.

instance of *authority*. This is to say that it has superior knowledge about a subject and consequently has something to teach the reader.[5]

Secondly, a text contains *information*. What Gadamer implies by this is that the content of the text is meaningful in some way. Thus the superior knowledge that it holds is meaningful and informative for the reader's life.

Thirdly, a text contains the *truth*. This last point is clearly contentious – a point that we will discuss later.[6]

Here, however, we can take note of the distinct direction Gadamer is taking. The earlier Romantics were not interested in the authority, information or truth of the text. They were concerned with the author's expression contained in the words, his originality, individuality, his thoughts and feelings. It is hard not to note the influence of a central critic of Romantic hermeneutics upon Gadamer, namely William Dray.[7]

A highly influential figure, Dray had criticised Dilthey along with other modern Romantics. He disapproved of what he considered to be their overly strong emphasis upon *context*. Whether this context was psychological – the feelings and intentions of the author – or indeed whether it was historical, this emphasis was, he believed, misplaced. In pursuing this *contextualisation*, Dray accused the Romantics of shifting the emphasis of the text *away* from the actual *meaning* of the text itself. Gadamer's emphasis upon the features of the *content* of the text, rather than its context, seeks to redress the imbalance perceived by Dray.[8]

Although shifting his notion of the principal features of the text, Gadamer builds upon the notion of grammatical interpretation inherited from the Romantics. He makes the same assumption which they did, namely that the text is a unity. The individual words, sentences, paragraphs and chapters all add together to convey the same unitary meaning.[9] The unified nature of the text is what permits the grammatical process of interpretation to occur. We can go into more detail here of its nature. A text's unity means that the parts and the whole, as we have said, fuse into a single meaning. This allows for a tracking back and forth between the parts and the whole in order to interpret a text. This tracking back and forth is known as the *hermeneutic circle*. Consider, for example, if I want to understand an unfamiliar passage of writing, I can read the

[5] Gadamer 1989: 277–85.
[6] Gadamer 1989: 277–85.
[7] Gadamer 1989: 174–231.
[8] See Dray 1959.
[9] Gadamer 1989: 265 (note, this assumption is challenged by later thinkers like Derrida).

preceding and succeeding passages. Through viewing the continuity of the narrative I can apprehend the meaning of a finite passage. Imagine, for heuristic sake, the contrary. If one were to select a few paragraphs from a favourite novel and read them individually, they would appear arbitrary and fail to make any sense. In order to understand one needs to have followed the unitary narrative. Moreover, certain ideas might only make sense once one has finished reading a book. At that point, one might wish to re-visit certain passages in order to revise their meaning in the light of completion. Obviously, when one embarks upon studying more complex texts, like theological or philosophical works, this process of reading the parts through the whole and vice versa becomes a cornerstone.

The Interpreter

To understand the second dimension of Gadamer's hermeneutics, we need to examine his notion of *the interpreter*. Gadamer had a very singular idea about the composition of the interpreter and further, that an elucidation of this was crucial to any serious hermeneutic theory. His ideas about the subject performing the act of interpretation, like his views about human beings in general, were informed by his teacher, Heidegger. Let us take a brief detour into Heidegger's keynotes on this topic and then see how Gadamer has borrowed them.

In common with many other German philosophers, Heidegger as we have seen, was interested in the nature of the existence of things in the world – the 'ontological'. Many German philosophers wanted, in particular, to capture the unique properties of *human* existence. According to their various views about this, philosophers have been characterised in distinct ways, as *idealists*, considering that human existence is a product of mind – that is the whole of the world we see around us is a product of either our own or a creator's (God's) mind. The infamous Hegel, for example, thought that human existence was the product of a 'super-historical' Mind. Other philosophers have been classified as materialists. Marx, for instance, believed that the essential nature of the human world was material; a product of socio-economic activity. Heidegger's view of human existence was special, too. For Heidegger, the most important point about us is that we are beings that exist in time.[10]

The idea that we exist in time seems at first to be both an obvious and an odd one. Heidegger, however, means two principal things by this. First,

---

[10] Gadamer 1989: 265–71.

we are distinct from physical objects. These are mere 'things' and are not interesting or ontologically significant. These physical objects have mere '*existence*'. The nature of humans is that we exist in time; we are temporal beings and have 'lives'. Thus, beyond mere existence, we have actual '*being*', which, as we have already seen, Heidegger refers to as *dasein*.[11]

The second point about human 'existence' as beings in time is that we are historical. You might think that this is an obvious point, too, but consider this view against alternative conceptions of humans. First, Heidegger counters popular Enlightenment notions about humans having a natural, psychological essence, which is implicitly a-historical. Secondly, Heidegger counters the Christian notion of human beings as inheritors of a soul from God. Finally, he also contradicts the Romantic idea, which emphasises the special uniqueness of individuals. Heidegger is not interested in forms of individual expression or indeed the accompanying fascination with genius that the Romantics held.[12]

For Heidegger, we are all historical beings. We are constituted by our own historical period and its inherited traditions. Note here, that for Heidegger to say that we are historical, means something very specific. He means that our human being is constituted by an inherited tradition. Again, we can contrast this with Enlightenment ideas of human beings, wherein, we would be regarded as consisting of certain a-historical processes, responses, psychological attributes. These would be independent of whatever era we happened to be born in. This notion of our being as constituted by an inherited tradition makes Heidegger a *conservative*.[13]

In sum, for Heidegger, human being consists of 'being in time'. On the one hand, this means that we are temporal, we are human lives. On the other hand, this entails that we are constituted by our historical time, which for the conservative Heidegger means a historical tradition.

Gadamer is influenced by Heidegger in his particular construction of the concept of the hermeneutic interpreter. To be precise, Gadamer borrows the notion of human beings as determined by their historically transmitted tradition. The interpreter in the hermeneutic process is thereby constituted by this historical tradition. What this means is that an interpreter approaches a text from a substantive perspective. He or she *already* holds certain kinds of knowledge and particular cultural norms and practices. He or she does not, therefore, conform to the majority of

[11] Gadamer 1989: 265–71.
[12] Gadamer 1989: 265–71.
[13] Gadamer 1989: 265–71.

the Enlightenment ideals or their contemporary scientific counterparts, which hold that a person wishing to gain understanding or knowledge should confront his or her subject matter with no prior views at all. They should indeed be free from any kind of assumptions. In science, the human being wishing to gain knowledge would conduct experiments and perform various kinds of controlled observations. These procedures for the gaining of knowledge would not be subject to the vagaries of history or tradition, but would hold true for all time. In scientific theory, if one repeated experiments at any moment during the course of history one would arrive at the same knowledge of the subject matter. In contrast, Gadamer believes that we, who want to gain understanding, are a product of our historically transmitted tradition. Thus, our ability to approach subject matter is, in turn, dependent upon, and constructed by this tradition.[14]

Gadamer's views are clearly deeply conservative – in the philosophical, indeed in the ontological, sense of the term. His emphasis is upon the past and upon tradition for the transmission of the possibility of understanding. Note that this is quite a different (although sometimes interconnected) notion from Conservatism as a political faction.[15]

Let us now progress to witness how these views held by Heidegger and transmitted to Gadamer affected the practice of hermeneutics itself. Gadamer offered a detailed notion of the interpreter and argued that the interpreter's 'mind' was formed by historical tradition. This entailed a special feature, for which Gadamer took a term from Heidegger, namely, the *forestructure* of understanding.[16]

The forestructure of understanding is really a fleshed-out way of conceptualising the process of understanding against scientific and earlier Enlightenment views. In contrast to these, which held that the mind was a *tabla rasa*, a blank slate for receiving sense impressions and knowledge, Gadamer held that there had to be *something* in your mind for it to understand anything at all.

This forestructure of understanding consisted in the main in what Gadamer unashamedly referred to as *prejudice*. Prejudice denotes our historically inherited sets of ideas, beliefs and notions. It does not mean for Gadamer what you and I would ordinarily take it to mean, that is a set

---

[14] Gadamer 1989: 291–300.

[15] Gadamer, however, was also a political conservative, albeit a rather quieter one than his mentor Heidegger.

[16] Gadamer 1989: 265–300.

of entrenched dislikes and ill-informed beliefs, for example, misogyny or racism. For Gadamer, prejudice means *pre-suppositions*. For example, we might have pre-suppositions about the importance of Shakespeare as a writer. These would be inherited from our English historical tradition. Whereas, from a scientific perspective, these views might be seen to interfere with a genuine *objective* view of Shakespeare's Macbeth or Hamlet, Gadamer believes that they constitute both the richness and indeed the very possibility of our understanding this literature at all.[17]

Prejudice, understood in Gadamer's sense is a positive and not a negative notion. Note, however, it contrasts with Romantic ideas. It is not based upon any individual, expressive, psychological or personal attribute but upon one's historically transmitted tradition. It is, therefore, not at all unique to an individual.

The Process of Understanding

The process of understanding can be thought of as a coming together of the text and the interpreter. We have examined Gadamer's view of the distinct nature of the text and the interpreter; we can now look at this process of the 'fusion' of the two: that is the act of acquiring understanding.

According to Gadamer, it is through the deployment of our inherited tradition upon the subject matter to be understood that allows us to gain understanding of its meaning. But how is this? How is it that, just because I have inherited preconceptions about the way things are, handed down to me through generations past, I can understand the meaning of a text? Gadamer's answer to this is that preconceptions allow us to *project* various meanings onto the subject matter to be understood. We can thereby throw outwards an anticipated meaning. Gadamer refers to this set of projected meanings as the *forestructure* of understanding.[18]

Gadamer then believes that this forestructure allows, us to understand because it allows us into the hermeneutic circle of meaning. Remember, that for hermeneuticists, no understanding can occur unless we can gain access to the hermeneutic circle, admittance to tracking back and forth between the parts and the whole of a text, constantly projecting and revising our understanding.

This process of projecting our inherited pre-suppositions, our prejudices, onto the text, and revising our understanding through the

---

[17] Gadamer 1989: 265–300.
[18] Gadamer 1989: 265–300.

hermeneutic circle, is known by Gadamer as the *fusion of horizons*. It is known thus, as it represents the contemporary interpreter fusing his or her inherited prejudices with the meanings from the past. Interpretation is thus an act of fusing past and present outlooks. The present historical 'horizon' meets the 'horizon' from the past.[19]

Notice how radically distinct from a scientific procedure this is, wherein our present prejudices would be seen to jeopardise a genuinely objective understanding. Notice, too, how different Gadamer's idea of understanding is from that of the Romantics. Understanding is not the recovery of an original meaning. For example, in contrast to Schleiermacher, we are not trying to understand what the author originally intended. Furthermore, Gadamer's view of historical understanding differs from Dilthey's. He believed that we needed to recover the original historical meaning of an act or event – imagine ourselves to be Napoleon or recover the meaning of a historical event as experienced and understood by those in the past. For Gadamer, understanding creates a 'new' meaning – a fusion of the present with the past. Indeed, all understanding for Gadamer is this fusion.

You might well begin to detect a possible problem with Gadamer's idea. What if, in the fore-structure of understanding, I project a 'wrong' meaning forwards onto the text. What, for example, if I read Plato's *Republic* believing that this is a text about art? Or worse still, if with no Classical education at all, I think that this is a text about modern republics? Surely then, the fusion of horizons will generate a defective interpretation?

Gadamer's answer to this appears to be that the hermeneutic circle will reveal such misconceptions. The hermeneutic circle consists in a unity of meaning. If I throw out a false presupposition, I cannot attain a unity of meaning. As I track back and forth between presupposition and text, the two will simply not 'fit': there will be a clash, a disunity. For instance, in reading *The Republic* expecting a discussion of ancient art, I will be confronted with the subject matter of ancient politics. I will thus have to revise my presupposition about *The Republic* and approach the text again with a new presupposition. Likewise, if I begin to read expecting a discussion of *modern* republicanism, I will be confronted with Ancient ideas and references. Again, I will then need to revise my presuppositions.

[19] Gadamer 1989: 265–300.

One of the reasons Gadamer believes the hermeneutic circle will reveal any inconsistencies, is that the past and the present belong to the same cultural tradition. As modern Westerners reading texts from the Western past we form a *unity* with that past. We are, for example, inheritors of many Ancient- and Christian-derived notions and as such live in a unity with the classical and spiritual texts that we wish to comprehend. When the fusion of horizons occurs, we simply bring together meanings from the past with those of the present – which are themselves linked to that past.

It is important to remember that although past and present meanings are fused in a cultural unity, interpretation for Gadamer is an ongoing process. In contrast to the earlier hermeneuticists, interpretation is neither the repetition of a 'stock' doctrinal reading, nor the recovery of an original (past) meaning. Interpretation is the generation of a '*new*' reading, one in which *past* and *present* fuse: a *fusion* of past and present horizons.

This 'new' reading, which is a combination of past historical understanding and historical tradition as transmitted in the present, Gadamer believes, brings tradition into the present. It regenerates historical meaning. Gadamer refers to this regeneration of historical tradition as *effective history*. It contrasts dramatically with Romantic-derived notions of historical interpretation as the nostalgic recovery of original meaning, which he regarded as a form of *preservative history*. For example, much intellectual history is concerned with understanding the meaning of terms and texts in their past historical context. This is a distinct project from the Gadamerian one, and would be considered by Gadamer to be, in actual fact, an impossibility. The need for the fore-structure of understanding, which is a product of the present, would render any attempt at unearthing pure, uncontaminated, past meanings impossible.[20]

There is a final point to note about Gadamer's view of interpretation. If we recall, from Heidegger, Gadamer appropriates the idea that the nature of our existence is temporal, we are historical beings and lives lived in time. For Gadamer, still following his mentor, a further part of the nature of this existence of ours, is that we interpret the world around us. Interpretation is simply part and parcel of our existence. This is to say interpretation is neither a choice activity nor an external method that we

---

[20] Gadamer 1989: 300–12.

deploy on certain occasions: it is simply part of what we are. *Interpretation*, thus, for Gadamer, too, has *ontological* status.

Gadamer's original question, 'what are the conditions for understanding?' led to the reply that interpretation is the key to understanding. This answer led in turn to the further question of how interpretation occurs. The response to this became, in the final analysis, that interpretation is ontological. This series of answers to issues about understanding is in stark contrast to a scientific mentality. In science, knowing is something we do. Moreover, it always requires some expert training with a set of highly specialised techniques. Furthermore, it can be practiced correctly or incorrectly. Science has particular methodologies which we may or may not deploy. Hermeneutics, however, is simply part of our being.

The bedrock of Gadamer's hermeneutics is that it is highly distinct from empirical (scientific) approaches to the study of the humanities. His is the pursuit of meaning and his difference with other hermeneuticists lies in his appropriating certain Heideggerian principles. Foremost amongst these is the notion of historical Being. Our historical Being entails that understanding, any understanding – especially understanding of the meaning of the past – is possible and as we have seen it is indeed ontological.

Gadamer's rather complex notion of interpretation can be simplified in the following chart.

TEXT:	truth, authority, information, unity of meaning.
INTERPRETER	Historical tradition provides *prejudices* (pre-suppositions) which are the *fore-structure* of understanding.
INTERPRETATION:	Fusion of present prejudice and past historical meaning: *fusion of horizons*.

There is an enormous amount of research and debate about the subject of understanding and an accompanying plethora of competing views about how to gain knowledge. We do not have space here for a comparison of hermeneutics with the many other perspectives. However, due to the increasing omnipresence of science in the humanities, it is worth making a comparison between Gadamer's hermeneutics and some key features of scientific approaches to 'understanding'. We compare a few key points in the following table.

Science	Gadamer's Hermeneutics
1. explanation	understanding.
2. causality/general laws	meaning.
3. external description	fusion of 'internal' and '(internal)-external' historical traditions.
4. 'do' scientific observation	'being' – understanding past of what we are.

We can see the trend in Romantic and post-Romantic German philosophy continued in the work of both Heidegger and Gadamer. Thinkers like von Humboldt and Dilthey had emphasised the linguistic and historical nature of human beings and told that, because we were part of history, we were internal to it and therefore could understand it from within. Heidegger and Gadamer have taken this notion further. We have not simply *made* history and therefore do not simply need to understand something we have ourselves constructed. We actually *are* historical. Thus understanding history is simply understanding that which we are. Furthermore, understanding that which we are is indeed constitutive of our very historical being.

This shift towards ever greater identification between human beings and historical meaning is not however undisputed. To see how it is disputed we can go on to examine the work of Gadamer's greatest critic, Jürgen Habermas.

HABERMAS

Jürgen Habermas, the German philosopher born in 1929, although not a hermeneuticist as strictly defined, is, however, an important correspondent with and critic of Gadamer. Furthermore, interpretative dimensions played a significant role in his own Marxist-leaning work in critical theory.[21] Habermas has many of the same intellectual enemies as Gadamer, notably the intrusion of crude scientific methodologies into the humanities, positivism in particular. However, as the old saying goes, one's enemy's enemy is not necessarily one's friend. In spite of the civility and

---

[21] We will be discussing Habermas's distinctive contribution to continental philosophy of social science at length in Part III of this book, but for now we need to relay his objections to Gadamerian hermeneutics.

intellectual sophistication of their exchange, Habermas and Gadamer are from intellectually very divergent and politically antagonistic traditions.[22]

It is in *Zur Logik der Sozialwissenschaften* that Habermas first engages with Gadamer.[23] His interest is sparked by the fact that he and Gadamer have the same objection to the positivistic approaches contaminating all branches of the humanities beyond social science into political, geographical, cultural and historical studies. Habermas, following his German idealist and historical materialist tradition, objects to the attempt to dismantle what broadly can be conceived as philosophical humanism and replace it with a rather superficial and literalist attempt to equate the fields of enquiry into human meaning with the practices of the natural sciences. There are numerous objections throughout Habermas's work to positivism. However, from the richness of a cultural tradition steeped in appreciation of the value of the historical, one feature of positivism is perhaps the most obviously objectionable, namely, the positivistic attempts to obliterate historical context from any understanding. The historical situated-ness of social meaning is a feature that is as precious to Habermas as it is to Gadamer.[24]

If we first look at the similarities between Habermas and Gadamer we find many unequivocal, jointly held oppositions to positivism. First, like Gadamer, Habermas's philosophy is steeped in historical sensitivity. Second, they both focus upon the importance of meaning and understanding in their work rather than causal analysis and explanation. Thirdly, Gadamer's notion of the fusion of horizons parallels Habermas's emphasis on understanding requiring *consensus*. In fact, they both employ the notion of understanding as dialogue. This differs dramatically from the positivist notion of an objective *tabula rasa* mind gathering information about its source. Moreover, both Habermas and Gadamer perceive interpretative understanding itself as dialogue rather than attempting to *reconstitute* an original meaning as the Romantics did, or merely the *appropriation* of meaning as the legal tradition held.

Their equally striking differences however emerge and can be seen to derive from their oppositional political, moral, and in the final analysis, what can only be described as ontological disagreements. First and foremost, Habermas accepts that whilst consensus is the aimed for dialogical model of understanding, in practice, such a consensus may be

---

[22] See Mendelson 1979.
[23] Habermas 1982.
[24] Habermas 1982.

'systematically distorted'. Gadamer, Habermas argues, leaves no space for the influence of an *ideological* context to understanding. What happens, Habermas asks, if consensus in his language, or the fusion of horizons in Gadamer's, is not in fact a product of dialogue and mutual understanding but instead is distorted by coercion? Gadamer's inability to even perceive this possibility makes hermeneutics nothing more than a 'linguistic idealism', he accuses. It assumes there are no socio-political or economic contextual factors shaping understanding. Habermas writes:

An interpretative sociology that hypostatises language to the subject of forms of life and of tradition ties itself to the idealist presupposition that linguistically artic- ulated consciousness determines the material practice of life. But the objective framework of social action is not exhausted by the dimension of intersubjectively intended and symbolically transmitted meaning. The linguistic infrastructure of society is part of a complex that, however symbolically mediated, is also consti- tuted by the constraint of reality – by the constraint of outer nature that enters into procedures for technical mastery and by the constraint of inner nature reflected in the repressive character of social relations.[25]

A further critical point included in Habermas's objection raised above, is the possibility of ideology having an internally coercive effect upon the consensus of understanding reached. For instance, what if one were to interpret *Mein Kampf*? Gadamer assumes that a text contains the truth. However, a text might merely be a vehicle of ideology. Not only does Gadamer refuse to acknowledge possible 'external' sources of distortion of meaning, but even if we were reading in a free and un-coerced context, he refuses to consider that the text itself might be the product of coer- cion and hence itself be ideologically distorted. Habermas is insistent, however, that we cannot simply assume that a text will be an instance of the truth. A further possible source of coercion could be that of the inter- preter himself. The interpreter after-all might also have ideological ends in mind. The consensus of understanding reached then, although both read in an un-coercive context, and indeed containing a truthful con- tent, could nevertheless still be ideologically distorted by the reader. In short, Habermas warns us against the likelihood in real socio-economic conditions of ever being able to assume a completely ideologically unin- fluenced consensus of understanding.[26]

This criticism reflects a deep division between Habermas and Gadamer, which we will see more clearly in the next section when we

[25] Habermas 1982: 289.
[26] Habermas 1982.

examine the Hegelian-Marxist tradition of continental philosophy and the approaches to the humanities it spawned. However, we ought to note that Habermas's critique of Gadamer rests upon a notion of a division between the socio-economic and the linguistic realm. Moreover, from the idea that the former is somehow foundational, ideological distortion is thereby likely to occur in all forms of efforts at textual interpretation. Moreover, a central problem for Habermas is that ideological distortion is almost certain to be 'hidden'; not straightforwardly manifest within the interpretative process. We must, he warns, guard against apparently 'innocent' interpretations which are necessarily bound up with the power relations themselves hidden within our society.

Gadamer's response to Habermas centres on his belief that hermeneutics can deal with the unravelling of distorted meanings within the interpretative process. He also argues that his notion of prejudice plays a key role here. Prejudice he claims, parallels ideology. Habermas, however, is not satisfied with this reply, believing that ideology has a deeper political base in society and is not merely the prejudice of a socio-historical era.[27]

Hermeneutics relies on the same categories as our own language, Habermas argues. If distortion is in our language itself, then the interpretative process itself is convoluted and cannot escape no matter how much internal interpretative procedure is performed. It simply has no external reference through which to unravel itself. Imagine a patient undergoing psychoanalysis. His or her self-understanding will be distorted by their own interpretative process. Their own language, by which they try to understand their own pathologies, will itself be wrapped up in those self-same pathologies. In the case of the patient, Habermas's own view about how to try to escape from this private language of psychopathologic deformation is to go back to the early childhood trauma that caused the pathology, and then, through re-enacting this, alleviate the problem. Herein, he points out, simple interpretation is not enough. Through the interpretative process one cannot escape the distorted meanings. To unravel them, one needs instead to go back to the initial cause and trace the development of the misrepresentation. That is, one needs to step outside of hermeneutics and look to earlier causes.[28]

Gadamer believes that, although prejudices need on occasion to be analysed themselves, the hermeneutic process is sufficient to achieve this:

[27] Mueller-Vollmer 1985: 294–319.
[28] Habermas 1982.

it will unravel any problems and the dialogue between past and present will eventually result in understanding. For Gadamer it cannot, of course, be otherwise for understanding is ontological. There is no deeper hidden layer to our existence. Habermas rejects this view because he thinks that understanding and language rest upon deeper, distinctive foundations and that distortions can appear in understanding due to the effects of these hidden underlying factors. Herein lies a fundamental ontological difference between these two philosophers which rebounds upon their competing conceptions of hermeneutics.[29]

With Gadamer's hermeneutics we had moved a long way from our classical and Christian predecessors and oddly enough, with Habermas's critique, we have traversed somewhat the distance back towards them. For our earlier hermeneuticists interpretation lay in two forms of historical practice, the literal and the allegorical. In the biblical tradition, history was perceived either as the literal procession of facts as depicted in the Bible, or as God's work on humans symbolising deeper spiritual meaning. First the Romantics and then Heidegger and Gadamer have shifted away from a Christian religious notion of historical meaning, but have continued the thread of the early allegorists by imbuing history with a deeper, hidden, meaning. However, whereas the allegorists perceived a separation between God's meanings, embodied within history and accessible only through allegorical interpretation, Heidegger, and Gadamer following him, have united historical meaning and human beings. Interpretation no longer necessitates overcoming the allegorical divide because we are within that meaning: that meaning is within us. If religious, Heidegger and Gadamer would have united God and man.

A point of parallel with their Christian predecessors is through the Reformation thinkers. Here, Luther and Calvin stressed the need for self-understanding as a means to gain closeness with God through the Bible. Heidegger and Gadamer can almost be read as an extension of that project. The whole process of interpretation, especially in Heidegger's thought, is that it occurs through historical understanding which amounts to self-understanding. But whilst Calvin emphasised the need to gain self-understanding in order to go beyond oneself and open oneself to the greater meaning of God, Heidegger seems to equate the greater meaning with historical Being itself. Our self-understanding is less a vehicle to go beyond ourselves than to generate a truthful encounter.

[29] See Mendelson 1979.

It is hard not to regard the lack of a truly external 'other' as a defi-
ciency in Heidegger's, and relatedly Gadamer's work. It entails that we
human beings somehow hold the truth within ourselves and our historical Being. There is no sense of an external standard bearer – be it
God, a further religious entity, a realm of moral truth (as in Kant), or
a notion of moral truth internal to history but external to ourselves (as
with Hegel). This amounts to the notion that what we ought to be is
somehow within what we are. This lack of a discrepancy between ourselves and moral Being is the point of criticism that is taken up most
strongly by Habermas. Although far from religious, Habermas's critique
is both 'epistemological' and moral. He questions the validity of interpretation remaining internal to ourselves as being able to generate any
true meaning. Furthermore, although couched in Marxist and elements
of Idealist theory, his critique amounts to a suspicion of this cosy internal unity between humans and historical meaning. He sheds doubt as
to whether this internal interpretation of history can be correct. He is
worried that factors might distort our ability to interpret history, factors
which obscure moral truth and 'epistemological' truth. Habermas's critique of (Heidegger and) Gadamer amounts to a rejection of the notion
of a unity between man and history. Interpretation of historical meaning, therefore, has to go back to surmounting a divide. Habermas's critique takes hermeneutics back to its earlier convictions, for instance, to
Augustine's belief that there is a symbolic divide between God's truth
and man. For Augustine, this had to be surmounted through the use of
allegorical technique. For Habermas this divide between humans and the
truth has to be surmounted through critical technique as well as interpretative understanding. In short, in parallel to Augustine's allegorical
technique is Habermas' dialectical approach to the humanities.

CONTEMPORARY HERMENEUTICS

In spite of the accusation of shortcomings, Gadamer's hermeneutics
was a watershed in the development of the tradition; so much so in
fact that twentieth-century hermeneutics, aside from Habermasian critique, has not really moved beyond the influence of many of the central
points outlined in *Truth and Method*. Most of the subsequent developments reflect this work in various ways. We can organise responses to
Gadamer into three different kinds. First, there are those who, as we have
seen, generated a critical response, for example, Habermas. Second, however, there are those who attempt to incorporate Gadamer's philosophy

into the new theoretical approaches dominant during their time. For instance, both Ricoeur and Derrida combine ideas from the important European movements of structuralist and post-structuralist thought with many Gadamerian hermeneutic insights. Finally, there are those who appropriate many of Gadamer's philosophical points, either directly or in a mediated fashion. They then apply these to the actual practice of interpretation itself, for instance, the American anthropologist Clifford Geertz is highly indebted to continental hermeneutics.

In the following sections we will view Gadamer's influence, both in generating a series of new theoretical hermeneutical projects, and in being appropriated in order to be applied to practical hermeneutical goals.

Ricoeur

Paul Ricoeur, born in Valence in 1913, was one of the leading philosophers in post-war France. He was influenced by Husserl and Heidegger whom he read whilst a prisoner in Germany during World War II. His was a fairly mobile career as he held Chairs at the University of Strasbourg, the Sorbonne and later went to Nanterre and Louvain, combining his last position with a part-time chair at the University of Chicago. He completed various philosophical works on problems of the will and phenomenology in his early career, moving on to examine psychoanalysis, structuralism and theory of the text in his later years. Ricoeur was responsible for bringing structuralist thought into the hermeneutic tradition. His seminal text in this regard was the highly influential – *De L'Interpretation* (1965).[30]

Structuralism, as we all know, was a movement prevalent in France from the early 1960s. It was originated by Saussure in the discipline of linguistics and especially semiotics, and went on to be associated with authors like Roland Barthes, Levi Strauss, Louis Althusser, and beyond France with Edmund Leach and Noam Chomsky.

Structuralism itself was influenced by Durkheim's notion of the social fact. Structuralist linguistics had several important 'Durkheimian' features. First, it assumed a scientific model of language. That is to say, language, like other entities, was presumed to be an object that could be investigated scientifically. Secondly, language represented a closed system of elements and rules that comprised the production and social communication of meaning. Language was a 'rule governed social systems of

---

[30] See Ricoeur 1976.

signs'. Although structuralism accepted the element of temporality, it focussed upon a 'static' system as its central object of analysis. Structural-ists examined language as a closed system of signs without reference to an external absolute. Meaning was internal to an autonomous system of signs.

What these structuralist influences meant for Ricoeur's hermeneutics was quite distinctive. First, because meaning was about a systems of signs particular to linguistic communities, it was less about subjectivity. Thus, Ricoeur shifted his emphasis dramatically away from issues of the author or indeed the interpreter. Further, he was also less interested in history than the object of language as it confronted him in the present. Moreover, like scientists in the natural sciences, he shifted his concerns from philo-sophical speculation to issues of method. Hermeneutics became about textual exegesis rather than philosophical speculation. He explains: 'We mean by hermeneutics the theory of rules that govern an exegesis, that is to say, an interpretation of a particular text or collection of signs sus-ceptible of being considered as a text'.[31]

In spite of the omnipresent influence of structuralism, Ricoeur's work demonstrates the influence of Gadamer in several crucial ways. First, he follows Gadamer in de-psychologising the hermeneutics of the Roman-tics like Schleiermacher and Dilthey who focussed so much upon the author's subjectivity. Secondly, like Gadamer, interpretation ought not to involve the imposition of the interpreter's context upon text. Interpreta-tion is not simply the appropriation of the text to the present or to the author's interests (Ricoeur accuses Bultmann of this). Thirdly, Ricoeur deploys a metaphor of the 'arch' to represent the uniting of the past with the present, the text with the interpretation. This metaphor, however, never goes as far as the Gadamerian insight of the fusion of the horizons between past and present.[32]

Having incorporated structuralism into Gadamerian hermeneutics, there were large differences generated between Ricoeur and his prede-cessor. One of the main distinctions was, in fact, in terms of the metaphor of interpretation representing a dialogue. Ricoeur differed dramatically from Gadamer in this respect. Structuralism held that there was a great division between speech and writing or reading. For structuralists, writing is, in fact, a distinct entity. Although distinct, it is equi-primordial, which is to say, there is no claim that speech comes before writing, or vice versa. Furthermore, writing (and reading) neither imitates nor fixes speech. It

[31] Ricoeur 1976: 18.
[32] Ricoeur 1976.

is simply different. For Ricoeur therefore, Gadamer's metaphor of the reading and interpreting of a text being like a dialogue falls apart for Ricoeur. Reading simply cannot, for him, be like dialogue at all.[33]

A further disparity between Ricoeur and Gadamer consists in their notion of meaning. Whilst for Gadamer meaning was constituted by a fusion of past and present, for Ricoeur, following his structuralist premises, meaning was internal to a system of signs and symbols. That is to say, the structuralists had a non-referential notion of 'meaning in itself'. For Ricoeur therefore, to understand meaning was to unravel these internal signs: it was an 'objective', intrinsic explication of text.

Consequent upon the distant notions of meaning, interpretation did not proceed with reference to historical context, past or present. The historical emphasis of Gadamer was all but entirely removed by Ricoeur.

A third feature of difference arises from Ricoeur extending one of Gadamer's notions to new extremes. We have already seen that Ricoeur echoed Gadamer by removing the Romantic emphasis upon the author's subjectivity from the act of interpretation. The meaning of a text being internal to a system of signs entails that there is not merely a lack of reference to historical context but also to the subjectivity of the author. The text is considered autonomous. This shift from the earlier psychological emphasis to the 'structure' of language is, in Ricoeur's case, total. It leads to his most influential notoriety in hermeneutic thought, known colloquially as 'the death of the author'. For Ricoeur, language speaks for itself. It does not entail a relationship with an author. A 'dialogue' with a text does not entail a dialogue with the author. Ricoeur claims that when reading a text it is as if the author were dead.[34]

Ricoeur's implication that interpretation is independent from historical context, authorial subjectivity or intention and indeed the interpreter's own subjectivity or intentions, implies a radical autonomy for texts. Ricoeur expresses this point when he writes:

By 'autonomy' I understand the independence of the text with respect to the intention of the author, the situation of the work and the original reader.[35]

However, it should be noted that the text is not completely autonomous and context free. It forms a relationship with language, other texts and the world of literature.

[33] Ricoeur 1976; see also Ricoeur 1984.
[34] Ricoeur 1984.
[35] Ricoeur 1981: 165.

A final dimension to Ricoeur's differences from Gadamer results from his aspiration to science – typical of all the structuralists. In order to interpret texts, structuralists aspire to use an objective method. In fact, Ricoeur's hermeneutics having jettisoned issues of subjectivity and history then got on with the business of formulating this method of interpretation. First, Ricoeur considered that the process of interpretation entailed a process of deciphering from *manifest* to *latent* meaning. That is to say, from a kind of 'conscious' to a kind of 'unconscious' meaning. For instance, in psychoanalysis, the latent meaning is deciphered from the manifest meaning of signs and symbols in dreams, and so on. For Ricoeur, symbols came in two kinds. First, those that were univocal which held a single meaning, for example definite labels like concepts and signs in logic. Secondly, there were those that were equivocal which held multiple meanings. Instances of these are myths, where we have an integrated system of signs which have both a manifest and a latent, deeper meaning.[36]

Different kinds of interpretation of symbols according to Ricoeur could either maintain or destroy the symbol. On the one hand, there were those who, in accessing the deeper meanings, conserved the symbolic realm of latent meaning. For instance, Bultmann in demythologising uncovers hidden meanings, but he retains the surface and deeper meaning. On the other hand, there exist modes of interpretation which according to Ricoeur, destroy the symbol by assuming that surface meanings are false: mere illusion. For example, in Freudian psychoanalysis, Ricoeur considers that in uncovering the illusion generated by the symbol, the symbol is destroyed. As a consequence of these creative and destructive approaches, it is impossible to generate universal rules for a method of interpretation. There are just too many different kinds of interpretation. However, for Ricoeur, the issue of method remained central.

Derrida

Moving on from Ricoeur we turn to consider another theorist who is seminal in the appropriation of many of Gadamer's ideas. Jacques Derrida was born in 1930 in Algeria to French parentage. During his life he has founded the International College of Philosophy in Paris and the International Group for Research and Teaching of Philosophy in 1975. He was most recently attached to the Etudes des Hautes Etudes en Sciences

[36] See Palmer 1969: 43–5.

Sociales in Paris. From among his many well-regarded texts, for instance, *Speech and Phenomena* (1973), *Writing and Difference* (1978), *Margins of Philosophy* (1983) and *Spectres of Marx* (1994), it is perhaps for his seminal, *Of Grammatology* (1977), that most will remember him.

Derrida like Ricoeur develops a new theory of hermeneutics incorporating as well as challenging some of the profoundest insights of Gadamer. However, his new developments are so critical of Gadamer that his project might be viewed by some as a merely critical one, more akin to Habermas's critical engagement than to an actual development in hermeneutics itself. Derrida is often associated, in fact, with the rather macabre project of the destruction of the tradition of the humanities itself. Based upon a selective reading of both Nietzsche and Heidegger he proposed the notion of 'deconstructionism' and indeed he is renowned as the founding father of this somewhat negative hermeneutic practice. This can be perceived to be a form of 'post-structuralism', building upon some of the ideas latent within the structuralist movement but seeking to exaggerate and undermine many of its principles. Deconstructionism has had an impact upon textual interpretation and makes sweeping claims for the tradition of hermeneutics. We can view Derrida's work overall as a fusion of deconstructionism and hermeneutics.

Deconstructionism

Derrida's project is termed post-structuralist because of his wide ranging critique of structuralist linguistics, an especial target here, of course, being Saussure. The motivation propelling this critique is a far-reaching antagonism towards metaphysics and he claims that structuralism, in spite of perceiving itself as 'beyond' metaphysics, itself draws upon many implicitly metaphysical assumptions. It is in his *Of Grammatology* that many of his most striking post-structuralist, deconstructionist claims are made. For instance, it is here Derrida claims that in retaining natural oppositions like those between speech and writing, inner and outer, structuralism retains hidden metaphysical assumptions.[37]

Deconstructionism is more than merely a critique of structuralism however. It is actually an attack upon the entire humanist philosophical tradition. Influenced by Heidegger, there is an attempt to do away with notions like a universal human essence. Directed principally against Enlightenment philosophy, Derrida attacks Kantian concepts like that of human being's (ideal) 'autonomous' status versus the contingent fluctuations of

[37] Derrida 1976.

the merely physical world. However, if this leads us to suspect a sympathy with Hegel or Marx, we would be very much mistaken. These philosophers still represent deep humanistic concerns, like a focus upon subjectivity as of central philosophical interest; moreover the notion that human consciousness has a 'metaphysical' constituting role. Far more undermining to this tradition, however, is Derrida's rejection of the humanist ideal of progress, be it the scientific or rationalistic ideals of Enlightenment as the progressive culmination of man's knowledge, or the Hegelian-Marxist belief in the progressive power of history. Derrida's notion of the end of history is far removed from the Hegelian-Marxist ideal of the fulfilment of historical goals. The end of history for Derrida represents the failure of the western tradition of humanism. It is a radical disenchantment and loss of faith, both in any Christian or in any Enlightenment sense.

The impact of these gloomy beliefs is writ large in a form of deconstructionist hermeneutics. Just as Derrida seeks to criticise and undermine humanist philosophy he also attempts to dismantle hermeneutic assumptions and practices.

All the traditions of hermeneutics thus far mentioned have certain features in common. The tradition as a whole is driven towards understanding through interpretation. Its major focus is upon the reading of texts. In order to accomplish understanding, many forms of hermeneutics draw first, upon information about the subjectivity of the author (Romantic), or indeed that of the interpreter (phenomenology, Bultmann).

Secondly, interpretation occurs against a backdrop of certain assumptions, for instance, that of meaning in the text. The text, it is assumed, can refer to objects and incidents beyond itself (most hermeneuticists make this assumption). Moreover, the interpreter often draws upon the context of texts, which besides the subjectivity of the author includes reference to historical context. The historical tradition of the interpreter is also often considered key to formulating an exegesis (e.g., Gadamer).

Thirdly, hermeneutics assumes a unitary nature to the text. Fourthly, interpreters always seek the most comprehensive account possible. Finally, hermeneutics differs from science in that it accepts that interpretations are always indefinite in contrast to the certainty claimed by scientific factual accounts.

Derrida provides a thoroughgoing critique of hermeneutics. This is mainly based upon his rejection of metaphysical assumptions. He pinpoints those implicit within the tradition of hermeneutics. First, Derrida opposes the idea of interpretation proceeding through a focus upon subjectivity, be it that of the original author or indeed that of the interpreter.

Taking the structuralist point of the 'death of the author' further, Derrida dismantles the idea of any unified subjectivity as a metaphysical fiction.[38]

Secondly, Derrida attacks the notion of the assumption of meaning in a text. One feature of this assumption is an implication that the text can refer to things beyond itself. Structuralism had anticipated the demise of the notion of texts referring beyond themselves by assuming that meaning was internal to a system of signs only. Derrida takes this idea further and argues that not only can systems of signs not go beyond themselves to refer to objects in the world, but that they are internally meaningless anyway.

Derrida's rebuttal of the notion of texts referring beyond themselves undermines a further conception in hermeneutics. Any self-respecting hermeneuticist when interpreting a text would take into account the context of the text, especially the historical circumstances in which it was written and also often the historical situation of the interpreter too. Derrida closes down the debate about how these contexts should be applied to understanding by simply stating that these milieu cannot be known and are irrelevant because the text cannot refer beyond itself.[39]

Thirdly, precious to hermeneutists is the assumption of a unitary nature to a text. In this respect, hermeneutics echoes structuralism wherein an internal unified meaning was assumed. Derrida, however, criticised structuralism's assumption of an internally coherent system of signs; that a text was a symbolic unity. This, he argued, was merely a metaphysical assumption. Thus within hermeneutics, Derrida attacked the idea of an assumed unity to the text.[40]

A fourth point of criticism that comes under Derrida's scrutinising eye, is the hermeneutic desire for a comprehensive reading of a text. This is again a metaphysical desire for complete knowledge, and it is, he argues, subject to the same illusions as metaphysics more generally. Indeed, Derrida goes further and not only claims that a comprehensive reading is an illusionary desire but that it is impossible indeed to gain any reading at all.[41]

This leads us to his fifth point. Previous hermeneuticists accepted that there was no single correct meaning, but that, in contrast to science,

[38] Derrida 1976. See also Hoy 1985.
[39] Derrida 1976.
[40] Derrida 1976.
[41] Derrida 1976.

there are several possible interpretations of texts with varying degrees of richness. Derrida postulates that more than the possibility of multiple readings, there are, in fact, an infinite amount of possible interpretations. Thus, he concludes, total undecidability reigns. Interpretation is, in fact, impossible.[42]

Deconstructionism, however, claims to be more than just a critique. It alleges to be a new way of 'doing' interpretation. Deconstructionist hermeneutics, as the name suggests, deconstructs texts. It proceeds by way of unpicking particular isolated phrases, or sentences and through these demonstrates the fragmented (as opposed to the unified), the incoherent (as opposed to the coherent), nature of the text. Deconstructionist hermeneutics thereby points not only to the inability of the internally self referential text to point beyond itself, but indeed to its internal incoherence and, therefore, lack of any meaning at all.

To my mind, deconstructionism is the inevitable consequence of an overly self-referential form of hermeneutics, such as Gadamer's Heideggerian variant. Ricoeur had prevented such a collapse by combining Gadamer with structuralism, although this simply takes hermeneutics closer to its scientific opposite than is appropriate or indeed constructive. Derrida's project is interesting, less as an approach in itself, than, perhaps appropriately, as a theoretical game showing where a unitary form of hermeneutics ends.

More constructive in contemporary hermeneutic developments, however, is the critical engagement of Habermas, and, for very different sorts of reasons, the application of Gadamerian hermeneutics in the actual process of interpretation itself. It is, therefore, to this latter that we now turn. We shall view the application of hermeneutics in the contemporary discipline of anthropology. In so doing we shall move away, in part, from a theoretical account and see the details of a vivid interpretation itself.

Geertz

Continental hermeneutics has made a powerful contribution to the modern discipline of anthropology. Within this discipline, especially within cultural anthropology, the practice of interpretation is fundamental. The cultural anthropologist wants to understand the meaning of rituals, symbols and cultural forms from other, usually foreign, cultures. They have to

---

[42] Derrida 1976.

work out what the unusual and distinct activities that they observe might mean to other peoples.

The first point here is to discern what kind of a meaning an anthropologist might be interested in. For example, do we want to know what a tribe's ritual means to *them*, to *us*, or are we trying to find some *objective* meaning within it? That is, do cultural activities have a fixed meaning, which is independent of the people taking part and independent, too, of the observer, or does the meaning of an activity apply only to the members within it? We can turn to hermeneutic theory for different answers to this question.

A second issue is the 'method' of gaining access to the meaning of foreign cultural events. Here we need to turn to an example of the application of anthropological interpretation. The example we are going to analyse is one of the most colourful in anthropology, the work of the contemporary American anthropologist Clifford Geertz. Note, although an examination of Geertz involves moving beyond the *geographical* boundaries of the Continent, this instance of the application of hermeneutics is steeped in Continental concerns and Geertz's intellectual debt to the continental hermeneutical tradition is thoroughly acknowledged, especially in his theoretical essays.[43]

Clifford Geertz was born in 1926 and is currently Professor of Social Science at the Institute for Advanced Study in Princeton. He is the distinguished author of numerous books and articles including the famous, *The Interpretation of Cultures* – a classic of the twentieth century. In this he explores both the theory and practice of hermeneutics in the human sciences with especial reference to anthropology, and to the locations of Bali and Morocco. In particular, in this chapter, we shall examine two of Geertz's key essays, the first on *Thick Description* and the second on *The Balinese Cock Fight*.

Within anthropology, as with other disciplines, there are many competing preoccupations. Some practitioners are interested in the material elements of cultures, the use or function of cultural activities to a particular group. Some anthropologists indeed consider that much apparently cultural activity is simply a means to various ends, such as economic security or physical survival. Some anthropologists consider that cultural activity obeys universal laws, which of course renders the activity of interpretation useless – we do not need to interpret individual events if they are merely manifestations of certain universal laws! Geertz's approach to

43 See for example, Geertz 1973: 3–32; 1993: 3–35.

anthropology is particularly useful as an instance of hermeneutics as he is interested in meaning within culture. He does not consider cultural activity to be merely an instantiation of a particular law, nor indeed, does he believe that all cultural activity can be explained by economic or pragmatic concerns. Geertz considers that cultural activity is meaningful and this meaning cannot be accessed through the deployment of symbols. This is because meaning occurs in symbolic systems. He writes:

The concept of culture I espouse, and whose utility the essays below attempt to demonstrate, is essentially a semiotic one. Believing, with Max Weber, that man is an animal suspended in webs of significance he himself has spun, I take culture to be those webs, and the analysis of it to be therefore not an experimental science in search of law but an interpretative one in search of meaning.[44]

Geertz's approach is to gain access to this meaning through interpretation. He writes: 'societies like lives, contain their own interpretations. One has only to learn how to gain access to them'.[45] How he achieves this is, in his own language, through what he refers to as *Thick Description*. This can be contrasted with *Thin Description*. Thin description is the process of mere empirical observation, where we record data. For example, what kind of clothes and artefacts a people use. Thin description can be immensely rich in detail, but it remains 'thin'.[46] Geertz's own work offers passages of exemplification of thin description:

Cockfights are held in a ring about fifty feet square. Usually they begin toward late afternoon and run three or four hours until sunset. About nine or ten separate matches comprise a program. Each match is precisely like the others in general pattern: there is no main match, no connection between individual matches, no variation in the format and each is arranged on a completely ad hoc basis.[47]

Geertz records more empirical content in minute details.

Thick description, in contrast with thin description, occurs when the anthropologist attempts to uncover the *meaning* of empirical data. This involves first, the process of 'thin description' that is the detailed observation as outlined earlier. However, it then goes on to uncover the meaning of what is observed. Thick description is not hard science, not the collection of facts, but a process of increasingly refined debate. 'Analysis then, is sorting out the structures of signification . . . Doing ethnography

---

[44] Geertz 1973: 5.
[45] Geertz 1973: 453.
[46] Geertz 1973: 3–32.
[47] Geertz 1973: 421.

is like trying to read...a manuscript – foreign, faded, full of ellipses, incoherencies, suspicious emendations, and tremendous commentaries, but written not in conventionalised graphs of sound but in transient examples of shaped behaviour'.[48]

Geertz's task then becomes one of how to uncover meaning. If the meaning of social action is expressed in symbols then we need to see how symbols accomplish representation. There are in fact two ways in which a symbol can represent a meaning. First, it can occur through conscious representation. For example, a flag represents a country. We all know what meaning is intended with the flag and if we ourselves want to represent that country, we would consciously choose that particular flag. Another instance of the idea of the conscious representation of an idea, might, for example be the Republican representation of 'liberty'. This is the rather erotic image of a woman dressed in white and blue with her breast exposed and holding a torch.[49]

A second way in which meaning can be represented is through 'hidden symbols'. This is when individuals, groups or indeed whole societies are unaware of what exactly a particular symbol represents. One manner in which these hidden meanings can be represented is through unconscious symbols. For example, Freud would say that if one had a dream about flying or running up and down stairs, that this would represent unconscious sexual desire on the part of the dreamer. The dreamer, however, would not be aware and indeed would have great difficulty coming to be aware of what the dream symbols represented. In order to gain access to these meanings, they would need the help of the interpretation of psychoanalysis.[50]

Cultural meanings are often, according to Geertz, in the latter category. That is, they are often hidden or unconscious. For this reason, we need a 'method' of interpretation in order to understand them. Geertz goes on to depict his own particular approach.

Social action occurs through the cultural forms of symbols, and these, according to Geertz, are like a text. Geertz expresses this idea in the title of one of his well-known essays 'the world in a text'.[51] That is to say, society is like a text. We can capture this text through the process of first, thin description wherein we meticulously capture all the empirical

[48] Geertz 1973: 9–10.
[49] Geertz 1973: 9.
[50] Geertz 1988: 9–20.
[51] Geertz 1988: 25–48.

details of social action through observation. Second, we then employ thick description, that is the process of interpreting meaning.

As we have seen the idea of social action as a text was first used by Dilthey, in his highly influential, *Hermeneutics and the Human Sciences*, with respect to human history. The French Hermeneuticist, Ricoeur then took up this idea and applied it social science by saying that, just as history can be considered to be a grand text, then so can all kinds of social action.

For Geertz, once we construe social action as a text we can deploy the features of a text in order to gain access to meaning. Geertz makes the following propositions about a text. First, it has a unity of meaning. Geertz explains 'as in more familiar exercises in close reading, one can start anywhere in a culture's repertoire of forms and end up anywhere else. One can stay, as I have here, within a single, more or less bounded form, and circle steadily within it'.[52]

Second, all the symbols unite to form a consistent narrative of substantive content. For example, Geertz claims, in his *Balinese Cockfight*, that a Cockerel symbolises masculinity. The overall event of a cockfight displays even more details of masculinity – further concordant meanings unfold.

Third, a text is not viewed as a static object, a mere 'thing' like a book. A text is considered by Geertz, following Ricoeur, to be dynamic. One actively engages with it. In point of fact a text is only truly a text, that is to say it only has meaning when one engages with it and performs the act of reading it. A book lying unopened on a shelf has no meaning – it must be actively engaged with. To express this another way, we could say that meaning lies not in the static object or 'thing', but in the act of engagement between the reader and the text. We can go on to say that all meaning resides in engagement with symbolic forms and not simply in the symbolic forms themselves.

Social Action as a Text

Let us now examine the particular instance of the *Balinese Cockfight*. In the social action of the *Balinese Cockfight* the text formed therein consists in symbolic forms of meaning. First, the Cockerels themselves represent certain cultural meanings. Geertz claims that the cockerels symbolise, on the one hand, masculinity or the male ego: 'cocks are symbolic expressions or magnifications of their owner's self, the narcissistic male ego writ out in Aesopian terms'. On the other hand, they represent natural animality against socialised civilisation. He continues in the same passage

---

[52] Geertz 1973: 473.

to claim that 'because although it is true that cocks are symbolic expressions or magnifications of their owner's self, the narcissistic male ego writ out in Aesopian terms, they are also expressions – and rather more immediate ones – of what the Balinese regard as the direct inversion, aesthetically, morally, and metaphysically, of human status: *animality*'.[53] He explains, further, how even 'The language of everyday moralism is shot through, on the male side of it, with roosterish imagery. *Sabung*, the word for cock (and one which appears in inscriptions as early as A.D. 92) is used metaphorically to mean hero', 'warrior', 'champion', 'man of parts', 'political candidate', 'bachelor', 'lady-killer' or 'tough guy'.[54]

In contrast with other common anthropological positions, Geertz is claiming that cockerels are neither, on the one hand simply domestic animals for breeding chickens and eggs for food or trading – economic use – that would be a material or pragmatic explanation. Nor, on the other, are they purveyors of a social function, e.g., that the cockerels generate social bonds through the activity of cockfight, which in turn serves the function of preserving social unity – this would be a functionalist explanation. Geertz is claiming that Cockerels have a meaningful role. They generate meaning within a particular culture.

Secondly, besides having a particular meaning themselves, cockerels also unite to constitute a broader text, to form a narrative in fact. Whilst the individual cockerel symbolises masculinity, the overall cockfight represents the broader context of this masculinity and within it, many more details unfold. 'In the cockfight, man and beast, good and evil, ego and id, the creative power of aroused masculinity and the destructive power of loosened animality fuse in a bloody drama of hatred, cruelty, violence, and death'.[55]

Third, the meaning of the cockfight lies not within the 'object' of the cockfight itself but within the act of *engagement* of the Balinese with the cockerels.

Enacted and *re-enacted*, so far without end, the cockfight enables the Balinese, as read and reread . . . to see a dimension of his own subjectivity. As he watches fight after fight, *with the active watching* of an owner and bettor. . . . [56]

Through engaging in the activity of the cockfight, Balinese males generate cultural meaning. That is to say that without the activity of the fight, the

---

[53] Geertz 1973: 419 (my emphasis).
[54] Geertz 1973: 148 (my emphasis).
[55] Geertz 1973: 420–21.
[56] Geertz:1973: 450 (my emphasis).

cockerels would have no symbolic meaning, they would simply be domestic fowl. This symbolic activity extends to even the passing of cockerels between men, as Geertz explains: 'Now and then, to get a feel for another bird, a man will fiddle this way with someone else's cock for a while, but usually by moving around to squat in place behind it, rather than just having it passed across to him as though it were merely an animal'.[57]

Let us now look at the process of interpreting these cultural meanings. This consists in the anthropologist performing several tasks. First, he or she must undergo thin description, meticulously observe and record the social action, see for example Geertz's extensive notes on the cockfight in *The Interpretation of Cultures* spanning from page 425 to page 448.

Second, the anthropologist must perform *thick description*, that is the interpretation of meaning. Herein, the anthropologist employs the *hermeneutic circle*. They assume that the social action is a text and that a unity of meaning resides within this. Hereby the anthropologist gains access to the meaning of individual symbols in the context of the overall meaning. Pages 448–453 of *The Interpretation of Cultures* are a superb instance of Geertz's own employment of this technique.

Third, in thick description the anthropologist is sensitive to how the Balinese themselves engage with the cockerels and generate meaning. He/she is receptive to the way in which the Balinese create their own text. As Geertz explains, the cockfight's 'function, if you want to call it that, is interpretative: it is the Balinese reading of Balinese experience, a story they tell themselves about themselves'.[58]

Finally, the anthropologist's own role as an interpreter is interesting. He or she in fact creates a second text, an interpretative one, by engaging with the Balinese's engagement: 'The culture of a people is an ensemble of texts . . . which the anthropologist strains to read over the shoulders of those to whom they properly belong'.[59] The act of interpretation generates new, further meanings beyond the original Balinese ones.

We can see, therefore, that the text exists at several layers. First, there is the '*original*' *text* of nature itself. A group of cockerels behaving as nature dictates. Secondly, there is the text of the Balinese cockfight. The Balinese are not accessing what the cockerels are naturally doing but are projecting their own meaning onto the cockerels. 'Attending cockfights

---

[57] Geertz 1973: 419.
[58] Geertz 1973: 448.
[59] Geertz 1973: 452.

and participating in them is, for the Balinese, a kind of sentimental education. What he learns there is *his culture*'s ethos and his private sensibility (or anyway, certain aspects of them) look like when spelled out externally in a collective text'.[60] This engagement creates the *text of the Balinese cockfight* and all its cultural meanings.

The third text is the *anthropologist's text*. This is his or her view of what the Balinese are doing. This consists of three features: (a) The object of the Balinese cockfight; (b) the process of the anthropologist's engagement with the Balinese cockfight – his or her detailed observation and interpretation; (c) finally, it also consists in the anthropologist creating a new text – his or her version of what the Balinese are doing. This last point is significant. The anthropologist's text is *not* the Balinese's own view of what they are doing. In this way it is parallel with psychoanalytic interpretation whereby we do not ask the dreamer the meaning of their own dream. Rather, we interpret the hidden, unconscious meaning through the use of our own external theoretical apparatus. In fact, Geertz does use psychoanalysis, an 'external' theory from Western culture in order to interpret the Balinese.

Let us now place Geertz's views and practice within the context of hermeneutic debates. If we compare his mode of practicing interpretation with that of his forebear Gadamer, we see a remarkable continuity. First, Geertz, like Gadamer, deploys the fore-structure of understanding. Geertz uses ideas from psychoanalysis, that is from Western culture and projects these onto the Balinese cockfight. It is through these culturally inherited 'prejudices' that the anthropologist is enabled to understand.

Secondly, Geertz also deploys the fusion of horizons. Understanding occurs when the Western cultural presuppositions of psychoanalysis are fused with the activities of the Balinese cockfight.

Thirdly, in contrast to earlier Romantic hermeneuticists, Geertz is not pursuing internal meaning. He is not reconstructing what the Balinese think the cockerels and the fight mean. He is not deploying Schleiermachian 'empathy': there is no attempt to 'become' like the Balinese. In contrast, for Geertz like Gadamer, interpretation generates a new text – one which is neither Western psychoanalysis nor the Balinese cockfight but a fusion of the two. This is a new, second text.

In short, interpretation in Geertz consists in three elements that are Gadamerian. (1) the fore-structure of understanding; (2) the fusion of horizons; (3) the creation of a new text.

[60] Geertz 1973: 449.

Although many features of Geertz's work are similar to Gadamer's hermeneutic approach, it is important not to mask certain important differences. These are as follows. First, importantly Geertz is not looking at cultural forms within the same historical tradition. Gadamer, following Dilthey, was concerned with history as a unified narrative which is how he could perceive the present as a unity with the past: they belong to the same internally coherent tradition. Geertz, however, in contrast is employing interpretation across traditions. He is deploying ideas from the Western cultural inheritance to examine a non-Western society. From a Gadamerian perspective, we would need to question the validity of this approach.

Secondly, Geertz is examining social action as a text. Unlike Gadamer, his concern is not with a written document from the past, however, but with action conducted in the present. Some hermeneutic scholars, following Gadamer and Ricoeur, have aimed some well-focussed criticisms at this point.

These critics claim that social action differs from a text and make a number of distinct points about exactly how these discrepancies occur.[61] First, they claim that a text is a very special kind of cultural form and that it has a particular structure. They question whether social action has the same kind of structure. Further they argue that it is because of the structure of a text, its unity of meaning etc., that the hermeneutic process can occur. The hermeneutic circle requires this structure – that symbols unite to form a consistent meaning. Therefore, the argument goes, if social action cannot be assumed to have this structure, and indeed for them it cannot, then the hermeneutic circle fails and the whole process of hermeneutics becomes impossible.

Secondly, in contrast to a text, these same critics claim, social action occurs over time. Thirdly, social action differs in a further dimension. Action, they argue, involves *agency*. A text, in contrast is 'pre-planned'. Anything that is pre-planned in this way, like a script for actors, rules out the factor of agency.

We can, however, defend Geertz from these criticisms in the following way. First, there is nothing to say that social action in the form of rituals does not have the structure of unity in the same way that a text does.

---

[61] See, for example, the criticisms implicit in Ricoeur 1981: 197–222, who expresses the idea that speech and texts are different so that action and texts must be even more so; a point more explicitly formulated in Thompson 1981.

Indeed it could also be argued that not all texts have a unified structure of meaning, which is indeed the stance of deconstructionists like Derrida. Secondly, Geertz believes that a written text's meaning only exists in relationship to the engagement of the reader upon the text. That is, the meaning only exists when the text is read. This, just as with social action, is a process that occurs over time. Reading a text is also, therefore, a form of social action. The meaning generated in reading a text is dynamic, not static. His defence, therefore, is less to say that social action is like a text than that for a text to be meaningful at all it itself must be a form of social action.

Finally, in response to the critic's third point above, Geertz believes that a text's meaning only exists in relationship with the author and his or her engagement. This engagement necessarily involves agency. Reading is every bit as much the deployment of agency as any other form of social action. That is to say, when reading, we do not have a pre-planned notion of what we decipher the meaning to be. We are engaged with the process of interpreting the meaning according to our (sometimes spontaneous) understanding which occurs over time.

From Geertz's use of a text, we could defend him from his critics by saying that for him, it is less that social action is different from a text, than that for a text to be meaningful at all, it must be like a form of social action. This makes Geertz's hermeneutics important, so let us just look at his exact notion of a text.

1. A text consists of symbolic forms which are not static elements waiting to be discovered on the page. Rather, these symbolic forms only exist and have meaning in the act of engagement of reading.
2. Interpreting a text is an act of creation and creates a new second text.
3. Reading a text is a form of action – it occurs over time.

In spite of the difference of interpreting across cultures, Geertz's interpretative practice reveals many features of Gadamer's theory. In 'thick description' Geertz applies the approach of interpretation and by treating social action as a text he provides a powerful example of how to gain access via the hermeneutic circle to other cultural forms and meanings.

Geertz concurs with the general interpretative perspective that the tradition of hermeneutics is perhaps the most extreme opposite to scientific approaches to study. In his study of the Balinese cockfight he is not concerned with explanation or with finding causes. Moreover, he does not deploy a process of controlled observation in the manner of science. He believes that we are part of the human meanings being studied whether these are internal or indeed external to our own culture. Hermeneutics in Geertz's own words is 'a process of ever refined debate rather than a precise science'.[62]

[62] Geertz 1973.

PART II

THE TRADITION OF GENEALOGY

The second part of Continental Philosophy of the Social Sciences broaches a further tradition of approaching understanding in the humanities. Whereas hermeneutics was principally concerned with inter-pretation, albeit of texts, historical events or cultural items, our new cus-tom of analysis is concerned with a specific kind of *historical* approach to understanding. This is the tradition of 'genealogy'.

The form of historical analysis that we are referring to here as 'geneal-ogy' is one that originated with the highly unique German thinker Friedrich Nietzsche. It has been highly influential across the continental tradition of the human sciences. Most notably it has been taken up and deployed by the French intellectual Michel Foucault.

Genealogy has many differences from hermeneutics, both in its history, its intentions and in its practice. In contrast to hermeneutics, which is always seen as a rather conservative tradition, genealogy is often perceived to be quite radical. Hermeneutics seeks to conserve traditions and to form a unity with the past; genealogy challenges many traditional assumptions and encourages a rupture with the past. Hermeneutics on the one hand often looks with reverence to our ancestors, genealogy on the other hand seeks to de-legitimise many of our assumptions about our heritage.

Although this second tradition of approach to the study of the human-ities is one, named by its creators, as 'genealogy', it is important to note from the outset that the way Nietzsche and Foucault use the notion of genealogy is quite distinct from its ordinary usage. I guess what you or I would think of as genealogy is a kind of tracing of family histories. We would do a genealogy perhaps of the paternal or indeed the maternal strand of the family. However, genealogy for Nietzsche and Foucault is very different from tracing a family tree or 'pedigree'. In fact, Geuss writes "Giving a genealogy is for Nietzsche the exact reverse of what we might call 'tracing a pedigree'.[1] Nietzsche's particular use of genealogy will be outlined in detail below.

[1] Geuss 1999: 1.

6

The History of Genealogy

Nietzsche

We begin with an introduction to the history of the 'tradition' of genealogy and discuss this approach as first articulated by Friedrich Nietzsche. We introduce the life and works of this infamous German philosopher and take a brief look at his early works and his views about knowledge. We then depict genealogy as an approach to the humanities used to critically analyse Christianity. We see how it was developed by the influential post-structuralist thinkers in the social sciences, in particular Foucault, whom we focus upon in the later chapters of this section.

NIETZSCHE

Friedrich Wilhelm Nietzsche was born in Germany in 1844. He lived until 1900, although he spent his latter years in a condition of complete insanity. He was a highly successful classicist, awarded a Chair in Classics at the University of Basle in 1869 at the startlingly young age of twenty-five. He then proceeded to write a succession of immensely influential books from the 1870s onwards. The style of all these works is similar, namely, poetic, aphoristic, highly expressive and emotionally charged. It was in 1889 that Nietzsche became plagued by insanity which lasted for eleven years until his death at the turn of the century. Tragically, he was not recognised during his sane life and then was only later perceived to be one of the most important European thinkers ever. Infamously, the Nazis made fallacious use of his philosophy decades after his death.[1]

---

[1] Two good biographies are Lavrin 1971 and Safranski 2002.

Nietzsche's work has been interpreted in almost every conceivable fashion: by the Nazis as one of their proponents, by the communists as one of theirs; likewise he was appropriated by anarchists and in contrast by conservatives. He is perhaps the most widely interpreted and misinterpreted philosopher ever.[2]

His seminal works begin with the *Birth of Tragedy*[3] written in 1871– 2, and a whole spate of key texts such as *Thus Spoke Zarathustra*,[4] *Beyond Good and Evil*,[5] *Genealogy of Morals*,[6] *Will to Power*,[7] from the 1880s onwards. There is clear continuity of style and indeed content in Nietzsche's works; he is perhaps best conceived as a cultural philosopher – although such a title might be rather demeaning given that there is no formally recognised sub-discipline of cultural philosophy. His works embrace discussions of aesthetics, science, culture, knowledge, morals, religion and history. In spite of continuities, Nietzsche develops quite differing views during the course of his intellectual life. Conventionally, his ideas have been grouped into an early, middle and late period. *The Birth of Tragedy* and *Untimely Meditations*[8] representing the main contributions of his early period, and its concern with aesthetics and the Greek ideal of Tragedy. *Human All To Human*[9] marks the onset of the middle period which is marked by a greater sympathy towards science and a more sceptical attitude towards art.[10] Of his many later seminal works, *Thus Spoke Zarathustra*[11] and *Ecce Homo*[12] are perhaps the best known, both becoming even more prophetic in style.

One of Nietzsche's key concerns throughout his life in common with many 19th- and 20th-century German thinkers was the maintenance of the high standards of the Western cultural tradition. He wanted to see the continuance of the richness of European arts and humanities. However even as early as the nineteenth century he began to witness the regression of cultural sophistication: Europe, he lamented, was losing that which had

---

[2] See Golomb 1997 or Ascheim 1992 for more details of the relationship between Nietzsche's work and anti-Semitism.
[3] Nietzsche 1993.
[4] Nietzsche 1961.
[5] Nietzsche 1990.
[6] Nietzsche 1994.
[7] Nietzsche 1924.
[8] Nietzsche 1997.
[9] Nietzsche 1997.
[10] See Vattimo 2002 more details.
[11] Nietzsche 1961.
[12] Nietzsche 1979.

made it great. He was fascinated by Tragedy and the Pre-Socratics and, in his early works, he believed that central to European strength was its foundations in Early Greek culture. His hostilities were of his times, but one can extrapolate from them to see that he would be afflicted by an even deeper horror of the contemporary state of the arts and humanities.[13]

Although Nietzsche's discussions embrace many issues pertinent to a philosophy of human sciences, this is not, of course, a subject which he specifically addressed.[14] However, in his philosophical challenge to many of the conventions of his time, he set the foundation for later philosophers, the French in particular, to construct new continental approaches to an understanding of 'social science'. Nietzsche's own contributions were to ideas about art and science and the relationship between these, his early works being of particular significance in challenging many of the epistemological assumptions upon which social science is based. Furthermore, his immense challenge to understanding the nature of history paved the way for the later genealogical approach which underpinned much of Foucault's later works, and is a topic of mainstay in philosophy of social science.

EARLY NIETZSCHE

Before beginning to examine the specific legacy of genealogy, it is well to look at Nietzsche's earliest work which sets the framework for his later intellectual oeuvre. This is no easy matter as Nietzsche is so notoriously difficult to encapsulate but, whatever the subtler tenets of genre in his early ambitions, he is clearly within the humanist and Romantic camp: early Nietzsche is no advocate of science or reason. In fact, stemming as he does from an eminent career as a scholar of Ancient Greek he is crucial in bringing a profound humanism into modern parlance. Indeed, Nietzsche, like the hermeneuticists, brings the sophistication of an old Western legacy into modern concerns and his challenging ideas spawn a highly influential cluster of approaches to contemporary European philosophy of social science.

The humanist influence in Nietzsche's work should not be underestimated due to his own rebellion against many underlying facets of this

---

[13] Universities awash with scientific approaches to the humanities and the arts overtaken by commercial and popular culture.

[14] To view the aspects of his philosophy which are seen more broadly as influential in the Philosophy of Science and epistemology see Babette and Cohen 1999, 1 and 2.

tradition – it is one thing to be consciously critical of a tradition over which one has complete mastery, and quite another to dismiss or ignore a canon of which one has little knowledge. Thus, although, as we shall see, Nietzsche confronts many humanist assumptions and practices, he brings this tradition forth into contemporary debate.

Early Nietzsche is propelled by an immersion in early Greek tragedy, by a conviction that the more powerful elements of Greek culture had been erased by a Christian tradition that looked only to the harmonious beauty of the Classical. Moreover, he is moved by Schopenhauer, by his darkness, his pessimism and his views about the essential role of art in understanding. Finally, the influence of Wagner complements his assumptions about the importance of art in human life.

These influences converge in his key early work, the *Birth of Tragedy*.[15] Herein he discusses the notion that artistic experience is the only vehicle through which we can come to see, or know, absolute Truth. Thus, against the notion of reason and science as the chief cognitive apparatus in our culture, Nietzsche places art at the forefront. This leads to a non-scientific, non-rationalist view of knowledge even more profoundly than the hermeneutic one. For the hermeneutic tradition's practice of interpretation at least revolves either around language with some rational grounding there, or it centres upon historical knowledge which entails some reliance upon empirical evidence thus drawing parallels with science. To believe that artistic experience is the essence of knowing in the way that Nietzsche does, is to allow neither logic nor empirical evidence in any sense to be connected with the process of gaining knowledge.

More especially, Nietzsche purports that it is a particular kind of aesthetic experience which gives us a window onto the deeper levels of existence. He mentions two kinds of artistic experience which he refers to as the Apolline and the Dionysiac. The Apolline refers to a gentle, 'poetic' type of aesthetic. Nietzsche calls it a kind of lyricism, and it is harmonious, allegedly the one from which the Western ideal of classicism is founded. According to Nietzsche, the Apolline has some ability to provide knowledge. He writes: 'This is the Apollonian dream state, in which the daylight world is veiled and a new world, more distinct, *comprehensible* and affecting the other and yet more shadowy, is constantly reborn before our eyes'.[16] Nietzsche continues with his claim about knowledge when he

---

[15] Nietzsche 1993.
[16] Nietzsche 1993: 14.

writes that in the Apollonian state 'all shapes speak to us'.[17] They 'tell us' something, provide us with some knowledge in an 'intuitive' direct fashion. However, the Apolline is less effective as a way of knowing the truth than the Dionysiac, but avoids some of the extreme consequences of the latter.

Nietzsche's second aesthetic category is the notion of the Dionysiac. This stems from the Greek God Dionysius, or the Bacchus, the Greek God of wine. Through wine, intoxication and orgy the aesthetic experience of complete self-abandonment is attained. Nietzsche writes that: 'we must imagine him as he sinks down ... from the revelling chorus in Dionysiac drunkenness and mystical self-negation, as his own condition, his unity with the innermost core of the world'.[18] Dionysiac *aesthetic* experience refers to any kind of art which entails a similar kind of self abandonment, however, it is principally music that has the capacity for this according to Nietzsche: Wagner's operas were the prime example. Nietzsche illustrates this when he quotes the final lines of Wagner's *Tristan and Isolde.*

> *In the sea of rapture's*
> *Surging roll*
> *In the fragrant waves'*
> *Ringing sound*
> *In the world-breath's*
> *Wafting space*
> *To drown – to sink*
> *Unconscious – supreme joy!*[19]

Nietzsche believed that within the Dionysiac lay the revelation of truth. The notion of 'truth' used here by Nietzsche is somewhat metaphysical, in the sense that it refers not to the truth of everyday facts, be they historical or scientific, but the deeper meaning of existence. However, given that human life is part of that deeper meaning, and that human life is organised into society, it would be fair to include society as part of that deeper 'truth'.

Self-abandonment through aesthetic experience entails a breaking of the boundaries around existent things. As Nietzsche describes it: 'The ecstasy of the Dionysiac state, abolishing the habitual barriers and boundaries of existence'.[20] It also entails a state of 'loss of self', 'subjectivity

---

[17] Nietzsche 1993: 15.
[18] Nietzsche 1993: 18.
[19] Nietzsche 1993: 106.
[20] Nietzsche 1993: 39.

becomes a complete forgetting of the self'.[21] Nietzsche explains: 'This self is not that of the waking, empirically real man . . . but rather the sole, truly existing and eternal self that dwells at the basis of being, through whose depictions the lyric genius sees right through to the very basis of being'.[22] So Nietzsche argues that to gain true knowledge, we need to exist in this altered state of 'consciousness' – indeed one of complete abandonment.

Behind this idea that the truth can be gained only through self-abandonment is the notion that logical categories, be they theoretical or empirical – from scientific testing etc., separate the mind from 'primordial being'. Almost all theories of cognition entail the idea that to gain knowledge we need to employ boundaries. We need to select bounded objects to know, and apply bounded concepts that demarcate one object from another. Nietzsche rejects this bounded notion of cognition and rails against Socrates who believed that self-reflection, concepts and logical thought provide us with knowledge. For Nietzsche, boundaries entail mere delusion. Herein lies a huge contrast with scientific or logical approaches to knowledge, which rely on increasingly sophisticated boundaries and demarcations between concepts, etc. In order to understand anything meaningful at all, including anything meaningful about human life and society, not only are these forms of knowledge inadequate, but Nietzsche argues they are positively deluded.

In order to gain access to the truth we need to lose ourselves through wine and orgy like the ancients, or drown in aesthetic (mainly musical) experience: 'to anyone who abandons himself entirely to the impression of a symphony . . . music express[es] the inner essence'.[23] This kind of knowledge entails a loss of self, a loss of the boundaries between objects, a merging or uniting between self and other and extreme pleasure. A disadvantage of this way of gaining knowledge is that, in the loss of self, in this complete aesthetic abandonment, we glimpse the eternal truth, and then die. In its completeness, aesthetic abandonment leads to certain death.

Nietzsche's early works are, of course, not themselves engaged with a contemporary sub-discipline like the philosophy of social science. However, we know that it was Nietzsche who, especially in his later work, provided a philosophical framework upon which much contemporary

[21] Nietzsche 1993: 17.
[22] Nietzsche 1993: 30.
[23] Nietzsche 1993: 78–9.

continental philosophy of social science is built and although philosophers of social science acknowledge this debt, and to a certain extent explore this, they mainly focus upon the impact of later Nietzsche's works, whilst tending to overlook the influence of his earlier works. Yet this early work remains interesting for two sorts of reasons. First, the implication of Nietzsche's argument's about art as the means to discover the truth has profound epistemological implications. His is a complete rejection of most Western epistemological positions.

Certainly it represents a refusal to accept that science or logic are the best modes of gaining knowledge. The idea that either science or reason could in any sense provide a basis for understanding society is therefore a nonsense for early Nietzsche. His position represents perhaps one of the most extreme epistemological positions: the complete rejection of scientific methods or logical analysis ideas applied to understanding. However, his alternatives are not clear. On the one hand, the Apollonian might entail that literary or poetic works are our best modes of understanding: Homer is the ultimate window onto human social life. On the other hand, even the Apollonian is limited and we need music and the Dionysiac if we have an appetite for the 'deeper' truth. The Dionysiac, if taken literally, may not exactly provide the most practical approach to the humanities: one would hardly recommend to one's students to indulge in a drunken Dionysiac orgy in order to reveal the ultimate truth about their social world or a literary text – although this 'method' might prove quite popular with some. But the point that, truth about society, if it is to be known at all, can emerge only through intoxication in music, for example, is a welcome relief from the rather emotionally barren conceptions of knowledge that pervade the Anglo-American approaches to the study of society. After all, if humans themselves are emotional and aesthetic creatures, ought not their society to be in part intrinsically emotional and aesthetic in its construction and therefore beyond the reaches of the rather vulgar quantitative methods of which sociologists are so very fond. This leads to a second point.

Nietzsche's arguments about art as the means to acquire the truth entails a different notion of truth from that which those with an epistemological predisposition towards science might hold. His concept 'truth' whilst difficult to define, contains a distinct notion of 'meaning'. This concept 'meaning', in turn itself difficult to define, is one which is less about categorising and defining objects in the world, and more about understanding their purposes and dare I say, 'spiritual significance', or the nature of their existence for themselves. Whilst modern commentators might regard this notion of truth as less significant for studying natural objects (unless one ventures into a complex mysticism), it might be highly pertinent with regard to social and cultural 'objects' which contain meanings of this more 'purposeful' nature for human beings. Therefore, Nietzsche's early works provide an instance of an extreme critique of science and reason in social science, moreover they suggest that aesthetic forms of knowledge are more relevant to understanding, and this latter might be especially relevant to an object-like society with its distinct kind of meaning.

A third point is that Nietzsche's early works strongly indicate the importance of the Classical tradition to European thought. Nietzsche himself builds his ideas and criticisms of modern society upon Greek ideas thereby demonstrating a humanist leaning, albeit rather unconventional. His European philosophy is built on foundations that look to the past, and is concerned with the maintenance of the might of a particular kind of classicism rather than with notions of progress. Furthermore, in his early works at least, and diffused throughout his later works, he pointedly advocates within his philosophy a return to the past, to the aesthetics of Greek Tragedy.

The Genealogy of Morals

In his later years, Nietzsche's position matured and shifted somewhat. Scholars have categorised his works in various ways, and, as we have seen, some have argued for continuity throughout his thought whilst others have perceived strong ruptures in his early, middle and later periods. Without getting sidetracked into specificities, it would be fair to say that there are clear patterns of continuity of subject matter, approach to philosophy and indeed mood in his work. Nietzsche acknowledges these himself and it is worth quoting at length his own insistence upon continuity in his thought. He writes, at the beginning of *Genealogy of Morals:*

This book was begun in Sorrento during the winter when it was given to me to pause as a wanderer pauses and look back across the broad and dangerous country my spirit had traversed up to that time. This was in the winter of 1876–77; *the ideas themselves are older. They were already in essentials the same ideas that I take up again in the present treatises* – let us hope that the long interval has done them good. They have become riper, clearer, stronger, more perfect! That I still cleave to them today, however, that they have become in the meantime more and more firmly attached to one another, indeed, entwined and interlaced with one another, strengthens my joyful assurance that they might have arisen in me from the first not as isolated, capricious, or sporadic things but from the common root, from a fundamental will of knowledge . . . .[24]

However, as Nietzsche matured, he clearly moved away from his Wagnerian Romanticism and developed highly specific attitudes towards social, historical and political concerns. In his middle works, he advocated science, less as an opposition to art and more as an extension of it.[25] However, his ambiguity towards science oscillated throughout his life. He also shifted concerns, from an early preoccupation with aesthetics to a later interest in more historical issues. In fact, in his later works, his analysis of his own culture and society was from a historical mode of analysis rather than an aesthetic or scientific one.

Genealogy of Morals is one of Nietzsche's principal later works which he describes as developing out of *Human All to Human*. As he writes at the beginning of the text, 'My ideas on the origin of our moral prejudices – for this is the subject of this polemic – received their first, brief, and provisional expression in the collection of aphorisms that bears the title *Human, All to Human. A Book for Free Spirits*'.[26] This text, *Genealogy of Morals* which is central to any understanding of European philosophy of social science, is a critical study of the central moral, cultural tenet of nineteenth Century Germany. It is founded upon a new philosophical framework that shook European tradition and provided a stepping stone for many later challenges to our most valued philosophical assumptions. Nietzsche's critical analysis of Christianity is centred around the development of a specific historical outlook, we could say, a 'philosophy of history', which can be described as a form of 'genealogy', although we must take care to note Nietzsche's very distinct use of this term.

Genealogy has sometimes been understood rather poorly,[27] whilst at other times scholars have encapsulated it rather well.[28] All commentators have, however, agreed that genealogy is a *particular* approach to historical analysis, but some have characterised it, and its later twentieth-century derivations, as a 'critical theory'. Although Nietzsche in his own genealogy is, in part, implicitly critical of the social norms and cultural values of his day we should demarcate genealogy strongly from the tradition

---

[24] Nietzsche 1994 (my emphasis).
[25] See for instance *Human All to Human*, Nietzsche 1997.
[26] Nietzsche 1994.
[27] See Habermas 1971, who actually attributes the opposite meaning to that which Nietzsche intended.
[28] See Nehamas 1985.

of Critical Theory. Whilst we could describe genealogy as a theoretical perspective on history that is likely to be implicitly critical, this is quite distinct from Critical Theory strictly defined, which employs a very specific notion of criticism. In fact, as Alexander Nehamas points out, for Nietzsche 'genealogy' is not some particular kind of special approach to history, albeit critical or otherwise, rather genealogy 'simply is history, correctly practiced'.[29]

Nietzsche's notion of genealogy is explicated and used in his study of Christian ethics. In *Genealogy of Morals* he challenges the alleged historical 'pedigree' of this dominant moral culture of his day. In place of Christianity's alleged pedigree Nietzsche posits the idea of a more random pattern to its establishment. Christianity's supremacy derives, he argues, not due to any inherent moral quality within it, but merely because of its success in various 'power struggles'.

We could compare this genealogical conception of Christianity to more mundane examples. Imagine, for instance, an individual with little talent who, by chance of circumstance, is employed in a prestigious post. The individual then due to this chance moment has a good career and explains his own success as being due to meritocratic factors. Another instance might be a businessman who, again with no inherent business acumen, purchases property in an area that for some unrelated reason proves popular, sores in price, and our lucky businessman's makes a fortune. He might then, and probably would, attribute his success to his pertinent insight. In both of these cases individuals account for their respective successes on the grounds of an origin in merit. The actual histories of chance, however, would undermine the alleged acumen, and indeed perhaps legitimacy, of the lucky individuals. But genealogy is more than just a history of 'luck'.[30]

Nietzsche's genealogy can be thought of philosophically in two ways. First, genealogy is *not* a theory of knowledge. It does not tell us how we can come to know the truth about the world. It is instead, *ontological*, it is a theory about the way the world actually *is*. In fact, it is a view about the world's nature as one of historical existence.

Secondly, Nietzsche's theory of genealogy can be characterised as nominalist. Nominalism, as we all know, is the opposite of essentialism. Essentialism means, like the name suggests, the 'essence' of something. That is

[29] Geuss 1999: 17.
[30] See Ansell-Pearson 1996 for a comparison of Nietzsche and Machiavelli encompassing the idea of a philosophy of luck.

to say, what something 'truly is', the property of an Object that is unchanging throughout time. A property which, without it, would cause an Object to no longer retain the same identity, or hold the same definition. For example, what is the essence of a chair? Is it that it is painted green or made of wood? Obviously not. Chairs can be any colour and can be made of metal or other materials. Is it that it has four legs? Well if a chair were to lose a leg, we would say that it was chair with a broken or lost leg: not that it wasn't a chair. Moreover, although chairs usually have four legs, some designs can have three and contemporary designs might consist in curved metal etc. with no legs. What might, therefore, be the essence of a chair? What might be its essential features that hold true for all chairs whatever style, culture or period in history? I guess one feature would be its purpose: you sit in it. We could say that the essential feature of a chair is that people sit in it. Now philosophers in general are not very interested in chairs. However, they are interested in the nature of human beings. A second example of essentialism, therefore, might derive from the answer to the question: 'what are the essential features of human beings'? Is it that they have two arms and two legs? Obviously not. Some people are born without limbs, or lose them in accidents or war, but they are still human beings. What might be the essential feature of a human being then? We could say having a 'self'. Moreover, having a self is perhaps the most important feature of being a human being.

Let us now look at essentialism as applied to an understanding of history. What might an essentialist view of history be? Bespeaking the obvious, history, if we are essentialists, would have an essence. Karl Marx, for example, thought this essence was dialectical materialism. The teleological unfolding of opposing material forces represented the essence of history. In particular, the development of tensions in the socio-economic realm constituted historical essence, all other features of history were either derived from, or incidental to this. The predominant view of history during Nietzsche's own times was, of course, also essentialist, not the secularised essentialism that held the passing of time to be driven by material forces, but the essentialism of a Christian view which held that history was the unfolding of God's intentions. In Nietzsche's own words: 'to interpret history to the glory of a divine reason, as the perpetual witness to a moral world order'.[31] Everything that happened throughout history whether good or evil, trivial or important, was the unfolding of a divine purpose. (Notice how the Christian view entails that history's

[31] Nietzsche 1994: 597.

essence is one of both purpose and meaning. Essentialist views often link the essential feature of an Object with its purpose or meaning.) In short, we can say that essentialism implies three features, an item has a definition, this is unchanging through time and is related to an inherent meaning or purpose.

Nominalism is the opposite of essentialism. In nominalism, we do not believe that things have an essence. There is, therefore, first, no definition to an item. We cannot define items but only describe them. It may be, for example, that at this moment in history, chairs are used for sitting on, but that in the past, they had other uses. We could not offer a single, unchanging definition of a chair.

Secondly, and relatedly, items have no aspect of their identity that is unchanging through time. For example, chairs may have changed through history. It could be that at a certain time in history, chairs were used to put food on, that is, as tables are used now. It could be that people always sat on the floor and then one day, somebody accidentally sat on a 'chair', and that for a time, this became a habit of others. The result was that most people started using chairs to sit on. In the present time, we would be so used to sitting on chairs that we would view them as having *always* existed. Moreover, we would see them as having been designed with that very purpose of sitting on them, that is, we would assume that they had originated in the past as a piece of furniture with the purpose of being a seat.

The idea that chairs originated with an essential feature (for sitting on) and that they have retained this same essential feature throughout history, would be an essentialist stance. In contrast, a history of the chair which claimed that it simply by accident came to be used for sitting on and that again, by some other accident it might be used for something else in the future, is a nominalist history. The essentialist believes that all things around us, be they chairs, habits, customs, events, institutions or human nature itself have an unchanging, essence to them. The nominalist believes that all these objects around us, even the human self, have no essential feature that is unchanging throughout history.

The final point is that the essentialist believes that this unchanging definition of an object captures its meaning. Nominalists, however, do not see meanings as fixed in this way. We cannot define an item with four legs and a back that we sit on as 'chair' because it might at some point in the past have had a different meaning. Likewise, the human self might have meant something quite distinct from the liberal individualist meaning we attribute to it now. As a further example, institutions themselves, for instance the criminal systems might have had quite different meanings in the past.

Nietzsche's theory of genealogy is both ontological and nominalist. That is to say, he depicts the nature of existence as historical: all existent objects are historically formed. Moreover, history is nominalist. Throughout history items have no single definition, no unchanging nature, no fixed meaning. Everything exists today due to its formation in history, but this formation is a sequence of chance events, not the unfolding of an essence.

Now that we have some picture of Nietzsche's ontological nominalism, we can move on to see how he elaborated this philosophical position. As we mentioned above, Nietzsche developed the idea of nominalism in relation to a challenge to the dominant Christian viewpoint of his day.

The first point to note, before examining the specificities of Nietzsche's critique, is that any nominalist position is intrinsically a critique of Christianity as this holds to both moral and metaphysical essentialism. As we have already pointed out, Nietzsche correctly surmises that Christians perceive history as the unfolding of divine intention. God's meaning permeates every strand of existence. Furthermore, Christian essentialism contained a moral doctrine: an unchanging central definition and meaning to morality. That which was morally good could be *defined* around certain principles, such as, 'doing good, giving, relieving, helping, encouraging, consoling, praising, rewarding; by prescribing "love of the neighbour"...'.[32] For Nietzsche, Christian values centre around 'concepts such as selflessness, self-denial', that is to say 'the unegoistic',[33] and moreover, the unegoistic typically took the form of 'pity, self-abnegation, self-sacrifice...'.[34] The essence to Christian morality is of the unegoistic, perhaps 'pity' as its central doctrine. For Christians this is unchanging over time, whilst in the Ancient World, pity was to be offered to slaves, in the Northern industrial Europe of Nietzsche's own nineteenth century, pity came through charity to the poor.

Against this unchanging essentialist view, any nominalist stance is directed towards refuting the possibility of an unchanging, a-historical definition. That which is Good cannot be defined and does change. For the nominalist, therefore, there is no possibility for un-egoistic behaviour to always be, by definition, Good.

---

[32] Nietzsche 1994: 571.
[33] Nietzsche 1994: 524.
[34] Nietzsche 1994: 455.

In Nietzsche's nominalism it is important to note that he makes two points. First, as we have seen, there are no unchanging essences to things, be these institutions, practices or indeed, morals etc., these change over time. This first point being that items change over time. Secondly, not only, according to Nietzsche, do items change over time, indeed they are *constituted* over time. Institutions, practices and morals are formed by history. For example, the idea of goodness, according to Nietzsche, not only changes according to whether Rome or Christendom dominate, but the meaning of goodness is actually constituted by these historical events. Nietzsche writes 'only that which has no history has a definition'.[35] If a definition captures the essential meaning of something, and if there are no essential meanings because things are constituted through history and change over time, then there can only be definitions of things which exist outside of history. For Nietzsche, nothing exists outside of history so there are no essences to anything. Put positively, everything existent has a history and therefore no essence. Morality has no essence of 'pity', this is a historical construct.

Genealogy

In *Genealogy of Morals*, Nietzsche's critique of Christianity is highly detailed. It occurs in characteristically caustic, aphoristic style, swaying between pure poetry and (rather less pure) rancour. In part, it is simply descriptive, typically opinionated and verges at times towards mere subjective ranting – for example, his comments on Judaism.[36] However, there also lies within his unending prose, a serious, even visionary critique, construed through the development of his ontological nominalist history.

Genealogy in Nietzsche can be conceived of, as we have seen, as the opposite of what we sometimes think of as genealogy, namely, tracing something back to its origins – a family name or precious heirloom. If we say that finding the pure origins of something is 'tracing a pedigree' then we can say that Nietzsche's genealogy is the very opposite of 'tracing a pedigree'.[37]

---

[35] Nietzsche 1994.

[36] Nietzsche 1994.

[37] As we have noted, some influential readings of Nietzsche get genealogy completely wrong, for instance, Jurgen Habermas actually attributes completely the reverse view to Nietzsche that he intended.

When we trace a pedigree, we look for a single, continuous, unbroken line of descent to the unitary source of an item. For example, a king traces the origin of his birth through an unbroken line of aristocratic descent. Geuss summarises how conventional genealogy, 'starting from a singular origin; which is an actual source of that value; traces an unbroken line of succession from the origin to that item; by a series of steps that preserve whatever is in question'.[38]

This tracing of his pedigree legitimises the king's rule and power. As Geuss puts it, tracing a pedigree is 'in the interests of a positive valorisation of some item'.[39] Of the fact that Nietzsche is against such practices of tracing a pedigree, he leaves us in no doubt. He writes of the 'historians of morality':

But it is, unhappily, certain that the historical spirit itself is lacking in them, that precisely all the good spirits of history have left them in the lurch! As is the hallowed custom with philosophers, the thinking of all of them is by nature unhistorical; there is no doubt about that. The way they have bungled their moral genealogy comes to light at the very beginning, where there task is to *investigate the origin* of the concept and judgement "good". '*Originally*" – so they decree . . . .[40]

Against the idea of this being a genuine historical practice of searching for the origins of morality and being a trace of historical descent, Nietzsche argues that:

it is plain to me, first of all, that in this theory the source of the concept "good" has been sought and established in the wrong place: the judgement 'good" did not originate with those to whom goodness was shown! Rather it was 'the good" themselves, that is to say, the noble, powerful, high-stationed and high-minded, who felt and established themselves and their actions as good, that is, of the first rank, in contradistinction to all the low, low-minded, common and plebeian.[41]

The conventional historical tracing of a pedigree merely serves to legitimate the actions of the 'high-stationed'.

In contrast, if we want to investigate the history of morality, Nietzsche's genealogy is the correct historical practice. Herein we look first, not for a single unitary origin: the search for origins is misplaced because there is no natural stopping place. For example, if we trace back a particular king's ancestors we might find eight grandparents all from different family lines. These family lines multiply and contains all sorts of tenuous and

---

[38] Geuss 1999: 3.
[39] Geuss 1999: 3.
[40] Nietzsche 1994: 461 (my emphasis).
[41] Nietzsche 1994: 462.

non-aristocratic links. If we go further back we perhaps find criminals and lowlife. Instead, if we want to examine an item's history, we glance backwards for *multiple lines of descent*. When Nietzsche looks for the etymological roots of 'goodness', he mentions German, Ancient Iranian and Slavic, Greek, Latin, Aryan, Celtic and more specifically Gaelic threads of 'origin'.[42]

Second, these multiple beginnings reveal not a direct, but *indirect* paths through history which are hard to perceive. Nietzsche declares 'that with the Jews there begins the slave revolt in morality: that revolt which has a history of two thousand years behind it and which we no longer see . . . '. He asks 'But you do not comprehend this? You are incapable of seeing something that required two thousand years to achieve victory? – There is nothing to wonder at in that: all *protracted* things are hard to see, *to see whole*'.[43]

Third, these pathways traced in a Nietzschean genealogical history are not intact (not 'whole') but *fragmented* and *dislocated*. Nietzsche describes how in the development of Christian morality, various pathways branch off and rupture. For instance: 'One will have divined already how easily the priestly mode of valuation can *branch off* from the knightly-aristocratic and then develop into its opposite; this is particularly likely when the priestly caste and the warrior caste are in jealous opposition to one another'.[44] Furthermore, he notes how these pathways in the historical construction of moral concepts often contain oppositions in their ruptured lines. With somewhat embarrassing anti-Semitic feeling, he writes: 'it was the Jews who, with awe inspiring consistency, dared to *invert* the aristocratic value-equation (good = noble = powerful = beautiful = happy = beloved of God) and to hang on to this inversion with their teeth . . . saying "the wretched alone are the good; the poor, impotent, lowly alone are the good; the suffering, deprived, sick, ugly alone are pious, alone are blessed by God, blessedness is for them alone – and you, the powerful and noble, are on the contrary evil" . . . .[45]

Nietzsche's final point about tracing a pedigree is that it legitimised an item through tracing an unbroken 'pure' line of descent from a supposed origin. In contrast, his genealogy through showing the haphazard changes through time, often serves to *delegitimise* an item. Nietzsche writes, continuing his criticisms of Christian morality, that 'with the Jews

---

[42] Nietzsche 1994: 464–7.
[43] Nietzsche 1994: 470 (my emphasis).
[44] Nietzsche 1994: 469 (my emphasis).
[45] Nietzsche 1994: 470 (my emphasis).

there begins the slave revolt in morality'... 'from the trunk of that tree of vengefulness and hatred, Jewish hatred – the profoundest and sublimest kind of hatred, capable of creating ideals and reversing values, the like of which has never existed on earth before...'.[46] He demonstrates the lack of any historical pedigree to Christian values. However, it would be a mistake to construe this delegitimising as a wholehearted rejection of an item. Nietzsche goes on in this same passage to state: '...there grew something equally incomparable, a new love, the profoundest and sublimest kind of love – and from what other trunk could it have grown?'.[47] The point is that Nietzsche's genealogy delegitimizes an item *in so far* as its legitimacy is based upon the false pretence of a historical pedigree.

The later Foucault, following Nietzsche, articulates all these Nietzschean features when he writes: "The search for descent is not the erecting of foundations: on the contrary, it disturbs what was previously considered immobile; it fragments what was thought unified; it shows the heterogeneity of what was imagined consistent with itself".[48]

We can compare the four features we have outlined above of tracing a pedigree and Nietzsche's genealogy.

Tracing a Pedigree	Genealogy
1. A single line of descent.	Multiple lines of descent.
2. Continuous, unbroken line of descent	Indirect, fragmented, dislocated, lines of descent.
3. Single unitary origin of an item.	No single origin, you can keep going back.
4. Legitimising role.	De-legitimises an item.

The Nature of History

From Nietzsche's ontological nominalist position we have seen that he attributes the existence of everything in the human world to its constitution through history. All words, concepts, meanings, institutions, practices, roles, positions and forms of behaviour are historically generated. That is to say, expressed in more common philosophical language, that all 'identities' are derived from history.[49] The mode of this historical

---

[46] Nietzsche 1994: 470.
[47] Nietzsche 1994: 470.
[48] Foucault 1977: 147.
[49] Note Nietzsche's nominalist position does not necessarily entail a radical instability to the identity of items. Historical change is perceived by him to be slow – Christianity existed for over two thousand years. Thus, Nietzsche is not an advocate of an unstable, vastly, ever-changing world as some later post-modernists believe.

constitution is genealogical. Items are constructed through multiple, broken, dispersed lines of descent. But what drives these multiple lines of descent? How do some institutions survive and change whilst others collapse? Why do certain meanings exist today, whilst others have been lost or replaced? For example, why has Christianity which was the dominant culture in Europe for centuries faded so markedly today?

Nietzsche was particularly interested in the history of morality. The question of why particular codes of ethical conduct dominated whilst others were all but lost became a pivotal one in his work. He was troubled especially about why Ancient Greek values which had once permeated earlier 'European' society, had been replaced during the course of history by Christian moral codes. How, he pondered, had Christianity come to dominate European culture so successfully. What was the secret of this historically composed ethic with its purportedly absolute moral truths?

Power

Nietzsche's answer was a sceptical one. He writes: 'The Origin of the world of moral concepts was, like origin of everything great on earth, long and thoroughly doused with blood'.[50] Although Christianity represented itself as truth, kindness and the moral good, its dominance was in fact the result of bloodshed. This bloodshed in turn Nietzsche argued, resulted from *power* struggles. Nietzsche analysed Christian morality in a highly perceptive and original fashion. The first point he made was to suggest that although the Christian conception of goodness was apparently a noble one (and indeed might even be a noble one) it resulted in fact from the successful power struggle of certain marginalized groups, notably slaves. It was the journey to power of these socially excluded groups that involved violence.

The Christian notion of goodness, as we have seen, revolved, for Nietzsche, around the concept of 'pity'. Pity was integral to the way in which these marginalized people legitimised the fight for empowerment. Moving our attention from the particular to the universal (although somewhat contrary to Nietzsche's approach) we can see that power is, in fact, what Nietzsche believed drove the historical constitution of all identities be they institutions, ideologies, beliefs, practices, concepts or meanings.

[50] Nietzsche 1994: II. 6.

Power, it must be understood is not necessarily in normative terms negative. Indeed, it can either be positive or negative. For example, when Christians deny that a bloody power struggle accounted for the success of the constitution of their values this, Nietzsche claims, would simply be erroneous. However, the fact that power struggles generated their moral culture does not imply a judgement on Nietzsche's part that Christianity is therefore necessarily evil. He leaves open this issue. His is a thorough-going ontological not an ethical critique. As to the issue of the normative standing of Christian ethics, Nietzsche is notoriously ambiguous. He is, on the one hand, ruthlessly critical of Christian morals. On the other hand, he expresses great admiration for elements of Christian compassion. Power is, therefore, not a normative notion – it is neither intrinsically good nor bad.

Furthermore, power does not necessarily mean domination or oppression. Through power one can oppress and dominate others, one can also, however, constitute identities according to Nietzsche. As a consequence of power one can achieve great things.

Contingency

There is a further feature to the historical constitution of items, namely contingency. If power drives history, contingency shapes it. The development of ideas and institutions is according to our great German master, haphazard. Civilisations grow and change in a random, chaotic way. Our most precious items and truths do not develop according to a single will or indeed through rational planning. Moreover, history is not the unfolding of a religious or historical purpose – as many Christians might argue. Furthermore, Nietzsche's idea of contingency as the seed of historical growth counters previous sacred German views like Hegel's or Marx's.

Like power, contingency for Nietzsche is simply the way things are. It is not a normative concept – contingency can be negative – it allowed weakness to subvert nobility. However, contingency can also be positive. Through a pattern of random change the ethical ideal of compassion arose. Moving on to consider more contemporary examples, on the negative front we can see that contingency allowed Nazism to rise to power in the twentieth Century. However, in the self-same century it facilitated the development of schools, hospitals and the welfare state. Contingency like power, operates intrinsically neither for good nor for evil. It is simply ontological.

We have looked at several key terms to describe Nietzsche's genealogy. We began with the ideas of ontology and nominalism and have gone on to describe genealogy, power and contingency. There is an important issue which we have yet to focus upon in a more detailed fashion in order to complete our understanding of genealogy. We need to examine the historical constitution of the *meanings* of words and institutions.

Like everything else in the human world, meanings are constituted historically. This implies several points. Most notably, it entails that as everything in history changes, so, too, do meanings themselves. For example, the meaning of the notion 'good' changes. An Ancient Greek notion of Good would mean that which fulfilled the human end of happiness. An Ancient Roman notion of goodness would be the courageous dedication to the strength and autonomy of one's country. The Christian notion of goodness would revolve around charity, compassion and forgiveness.

These meanings of goodness change over history according to power and contingency, rather than because one is necessarily better through some criteria than another. The reason why goodness came to mean happiness in Ancient Greece was because a particular strand of Ancient Greek culture was successful in various power struggles and because of various multiple contingent factors. Likewise, Roman goodness meant courage for the liberty of the Patria because this culture of 'Virtue' became powerful in Rome, and Rome became powerful in the rest of the world. Thereafter, goodness meant 'pity' as Christianity succeeded in defeating the Ancients through contingency and power struggles.

The meaning of words, practices and institutions 'developed' not in a systematic, pre-ordained or unitary fashion. Meanings were constituted genealogically. They changed in a haphazard way, through multiple disconnected lines which cannot be traced back to any singular origin.

Finally, Nietzsche believes that meanings grow through a constellation. This means that new meanings do not *replace* old ones. The old ones resist the imposition of new meaning whilst the new meanings lie over the old. Items thereby accumulate layers of meaning. Goodness, for example, still retains the idea of virtue as courage for the Patria, the notion of human happiness or fulfilment as well as the Christian idea of compassion and pity.

Construction

It is a final and perhaps most important feature to note about genealogy. The genealogical process of history constructs items. For instance, Nietzsche explains how Christian morals grow from the trunk of hatred into something beautiful. Out of that which is antithetical to goodness comes something good. He writes:

from the trunk of that tree of vengefulness and hatred, Jewish hatred – the profoundest and sublimest kind of hatred, capable of creating ideals and reversing values, the like of which has never existed on earth before... there grew something equally incomparable, a new love, the profoundest and sublimest kind of love – and from what other trunk could it have grown?.[51]

Besides Nietzsche's love of paradox, and sense of the paradoxes within history, he employs the notion of historical process as constructing Christian moral goodness. In his genealogy he explains two of the threads in the development of the Christian concept of 'pity'. One 'origin' of pity was in a power struggle, the inversion of the lowly to become high. It was driven by hatred. Those who were the pitied resented the strong. Pity was a mechanism of their empowerment and legitimacy. Another movement in pity's history was its inversion as the Christian power struggle succeeded. Then it entailed the oppression of the new weak as the old weak became strong. Through pity the successful Christians subjugated and humiliated the new poor. They provide for them, with pity, whilst at the same time aggrandizing themselves and controlling, perhaps through charity, those they pitied. Within these dark pathways, however, Nietzsche acknowledges that the Christian pity of love thy neighbour has a sublime element. In Christianity, a genuine good has been constructed through historical process.

In summary, meanings are arrived at in the following way:

  i. – They are historically constituted – they change over time.
  ii. – They change through power and contingency
  iii. – They have fragmented disconnected multiple lines of descent.
  iv. – Meanings accumulate – history doesn't erase old meanings, it adds and transforms them.
  v. – Meanings are a synthesis or constellation of past and present.
  vi. – Old meanings remain and resist imposition of new meanings.
  vii. – History constructs items.

We need to address now what the significance of this genealogy is for the philosophy of social science. Nietzsche's thought, elaborated above, in fact contains many ideas that have influenced Continental philosophy of social science. Most notably perhaps his nominalism has led to a

[51] Nietzsche 1994: 470.

radical challenge to many social scientific epistemological assumptions. Furthermore, Nietzsche's view that society is historically constituted leads to the important point that society needs to be understood, if indeed it is to be understood at all, *historically*. This emphasis upon the importance of history, along with a focus upon the Christian and Ancient pasts, to my mind, places Nietzsche's approach to social analysis within the deeper humanist tradition. However, his nominalist history provides a challenge to the very tradition from which it emanates.

Followers of Nietzsche have been influenced to a greater or lesser degree either by his humanism or by his nominalism. Whilst projects like Derrida's deconstructionism take the latter nominalist element to an extreme, others, like Foucault, incorporate Nietzsche's humanistic with his nominalist elements. Indeed, Foucault, perhaps in spite of his own self image, and certainly in contrast to many of his later followers, maintained a high degree of humanism by studying society and its institutions historically, and by following through the threads of an image of the Ancient and Christian past and using this in order to understand the present.

More specifically, if we were to follow Nietzsche's genealogy as a mode of pursuing philosophy of social science, we would be of the persuasion that in order to understand anything in the humanities, be it a feature of society, a cultural artefact, an institution, code of practice, ideal, set of beliefs, a literary text, a system of moral conduct, a political system or the features of crime and punishment, we would need to do a genealogy. We would have to unravel meanings, trace back the power struggles and contingencies that have constituted each meaning in our society, be it a moral truth like 'pity', an institutional practice like punishment, a value like that of the importance of education. We must do genealogies of the sovereign and liberal power regimes of our own times, analyse the dominance of forms of popular culture in the music industry and in sport. These political and popular icons of our own society all stem from contingency and successful power struggles. We need to trace back the multiple, disconnected lines of descent and in so doing, we potentially de-legitimise these social features. For example, we de-legitimise sovereign right in so far as it is based upon the claim of a pure unbroken origin.

In short, everything that is a feature of our society has a genealogy. This isn't inherently good or bad, but it is, if we follow Nietzsche, the only correct way of doing 'social science'.

7

Theory of Genealogy

Foucault

Nietzsche himself is often neglected by continental philosophers of social science who all too frequently overlook the historical underpinnings of the modern discipline. However, the complexity and importance of the work of Nietzsche's disciple Michel Foucault cannot be fully appreciated without a familiarity with his German predecessor.

Michel Foucault has proved a pivotal figure for almost all branches of the twentieth and twenty-first century humanities. Sadly, his work, too, is often poorly depicted in sociology, human geography and cultural studies where Foucault is portrayed in scant theoretical style leading those from traditional disciplines to regard him as little more than a flippant peddler of some perfidious version of post-modernism. This would be a grave injustice as Foucault is perhaps the best-ever student of Nietzsche, appropriating a visionary analysis and applying it to his own society, with dramatic effect. Foucault was, in fact, the first person to correctly understand Nietzsche's genealogy and to apply it with new potency. If Nietzsche's genealogy was directed towards an understanding of nineteenth-century morals, Foucault appropriated this same nominalist, historical approach and deployed it in understanding the features of twentieth-century European society.

Michel Foucault, the flamboyant French philosopher with characteristic shaven head is familiar to us all. Born in Poitiers in 1926, he died in 1984. The Ecole Normale Superieure was where he studied from 1946 and he held various posts thereafter throughout France, most notably in the College de France from 1969. His early works were on mental illness and he worked within a Marxist and existentialist phenomenological framework. He soon moved on to more philosophical and historical analyses although he continued to depict madness, asylums and clinical medicine. Later he added to these studies further works in the social sciences and in theories of knowledge, which are often described by a form of analysis known as 'archaeology'. Later he moved on to a reckoning of the further institutions of reform, the prisons and the accompanying human sciences. His final works were on health, the body and sexuality.

For our discussion we are interested in seeing genealogy as an approach to the humanities and herein we are best informed by Foucault's later works. The most important of these are his studies of power and knowledge in the human sciences contained within his *Power/Knowledge*.[1] Moreover, his highly influential depiction of penal regimes in the notorious *Discipline and Punish* is invaluable.[2]

Foucault's later works were clearly genealogical. Consider, for example, his emphasis, on broken lines of descent in his studies on knowledge. He writes "It seemed to me that in certain empirical forms of knowledge like biology, political economy, psychiatry, medicine, etc. the rhythm of transformation doesn't follow the smooth, continuist schemas of development which are normally accepted".[3] Clearly this perspective is highly indebted to Nietzsche. Foucault was, in fact, not so much an original philosopher, as Nietzsche had been, but was, in fact, more of an applied sociologist and historian. What was original, however, was his in depth understanding of genealogy and his highly systematic appropriation and application of Nietzsche's ideas to the cultural context of his own times. This was nowhere more true than in the case of Foucault's studies on power/knowledge in the human sciences and in his analysis of the penal regime.

Before we study Foucault's use of genealogy, let us first pause to look at his own direct writings on the subject. In an essay entitled *Nietzsche, Genealogy and History*,[4] we see how a brand of German philosophy is taken up by this French mind preoccupied with the subject matter of the social sciences. First, Foucault begins in sympathy with Nietzsche's approach to a historical study of morality. He writes: '...it is obvious that...[it is]...wrong to follow...the history of morality in terms of a linear development'. He criticises those that followed essentialist

[1] Foucault 1980.
[2] Foucault 1979.
[3] Foucault 1980: 111–12.
[4] Foucault 1977.

historical assumptions and 'assumed that words had kept their meaning, that desires still pointed in a single direction, and that ideas retained their logic'. Historians such as these had 'ignored the fact that the world of speech and desires has known invasions, struggles, plundering, disguises, ploys.'.[5] Foucault continues in a Nietzschean vein to claim that 'genealogy does not oppose itself to history as the lofty and profound gaze of the philosophers might compare to the mole like perspective of the scholar; on the contrary, it rejects the metahistorical deployment of ideal significations and indefinite teleologies. It opposes itself to the search for "origins".'[6]

If the former demonstrates Foucault's commitment to a genealogical perspective, the following reveals the detail of his familiarity with and wholehearted embrace of Nietzsche's writings on the topic. Foucault denotes the different German terms for 'origin', although it is less this that interests us than the fact that his detailed search through Nietzsche's use of the term reveals an impressive knowledge of this philosopher's works. Foucault describes *Genealogy of Morals*, *The Gay Science*, *Human all to Human*[7] and *Untimely Meditations*[8] as expressing the development of Nietzsche's views about the search for 'origins' in history and his generation of the new genealogical perspective. But Foucault is not simply an erudite scholar of Nietzsche's works, and does not merely replicate his ideas. He takes them on board and uses them to argue how our most precious notions, like the self, the body, liberty, crime and punishment, feelings, instincts and indeed the truth itself can all be shown to have genealogies.[9]

Before examining Foucault's application of genealogy, it is as well to note that his selection of the features of a genealogical approach exactly mirrors Nietzsche's own. In a passage that it is worth quoting at length, he rails against Nietzsche's old enemy, historical essentialism.

Why does Nietzsche challenges the pursuit of the origin (*Ursprung*), at least on those occasions when he is truly a genealogist? First, because it is an attempt to capture the exact essence of things, their purest possibilities, and their carefully protected identities, because this search assumes the existence of immobile forms that precede the external world of accident and succession. This search is directed to 'that which was already there', 'the image of a primordial truth fully adequate

---

5 Foucault 1977: 139.
6 Foucault 1977: 140.
7 Foucault 1977: 140.
8 Foucault 1977: 164.
9 Foucault 1977: 140–64.

to its nature', and it necessitates the removal of every mask to ultimately disclose an original identity. However, if he listens to history, he finds there is "something altogether different" behind things: not a timeless and essential secret *but the secret that they have no essence* or that their essence was fabricated in a piecemeal fashion from alien forms.[10]

This passage contains the out and out rejection of essentialism, and moreover, many of precisely the same features in the historical practice of tracing origins that Nietzsche himself rejects.

Foucault moves on to identify further features that he rejects in 'conventional' essentialist history. The first one that echoes Nietzsche's analysis, is obviously the rejection of 'origins' themselves. This rejection is twofold. On the one hand, he rejects the notion of a *single* origin. He writes 'What is found at the historical beginning of things is not the inviolable identity of their origin; it is the dissension of other things. It is *disparity*'.[11] On the other hand, if origins are no longer to be considered as singular, but as disparate, then so too, they are no longer perceived as unitary. Foucault explains: 'The search for descent is not the erecting of foundations: on the contrary, it disturbs what was previously considered immobile; *it fragments* what was thought *unified*; it shows the *heterogeneity* of what was *imagined consistent with itself*?'.[12] As with Nietzsche's, in Foucault's genealogy origins are multiple and heterogeneous.

Secondly, as the origin is unmasked in genealogy, so too are the lines of descent themselves. Again, this rejection is twofold. On the one hand, there is the rejection of the idea of a single line of descent. Foucault expresses this as: 'we should not be deceived into thinking that this heritage is an acquisition, a possession that grows and solidifies; rather, it is an unstable assemblage of faults, fissures, and heterogeneous layers that threaten the fragile inheritor from within or from underneath'.[13] Lines of descent, like 'origins' are not singular but rather, heterogeneous. On the other hand, Foucault also refuses to accept descent-lines as continuous. In his words, he writes 'genealogy does not pretend to go back in time to restore an unbroken continuity that operates beyond the dispersion of forgotten things'.[14] Historical lines of descent are not continuous, but with 'fissures,' that is, discontinuous.

[10] Foucault 1977: 142 (my emphasis).
[11] Foucault 1977: 142.
[12] Foucault 1977: 146 (my emphasis).
[13] Foucault 1977: 146.
[14] Foucault 1977: 146.

A third feature of Foucault's genealogy is its role in delegitimising an item, in so far as an item's legitimacy is gained through its pretence to a pure line of descent. He explains:

> The forces operating in history... do not manifest the successive forms of a primordial intention and their attraction is not that of a conclusion, for they always appear through singular randomness of events. The inverse of the Christian world, spun entirely by a divine spider, and different from the world of the Greeks, divided between the realm of will and the great cosmic folly, the world of effective history knows only one kingdom, without providence or final cause, where there is only "the iron hand of necessity shaking the dice-box of chance"... The world we know is not this ultimately simple configuration where events are reduced to accentuate their essential traits, their final meaning, or their initial and final value.[15]

In point of fact, genealogy in revealing an item's multiple layers of descent 'threaten[s] the fragile inheritor from within or from underneath'.[16]

Just as Foucault followed Nietzsche in his belief that the pathways through history were multiple and without clear beginnings, so, too, did he account for the 'driving force' in history as being that of power. Foucault writes; 'developments may appear as a culmination, but they are merely episodes in a series of subjugations'.[17] He continues, 'In a sense, only a single drama is ever staged in this 'non-place", the endlessly repeated play of dominations'.[18] In fact, he argues, power struggles do not simply account for our past, but for our historically constituted present also. He explains 'Humanity does not gradually progress from combat to combat until it arrives at universal reciprocity, where the rule of law finally replaces warfare; humanity installs each of its violences in a system of rules and thus proceeds from domination to domination'.[19]

Foucault also follows Nietzsche in asserting that the two shaping forces to history are power, on the one hand, as we have just seen, and contingency on the other. He states: 'The forces operating in history are not controlled by destiny or regulative mechanisms, but respond to *haphazard* conflicts'.[20] We have already heard Foucault refer to 'the dice-box of

[15] Foucault 1977: 155.
[16] Foucault 1977: 146.
[17] Foucault 1977: 148.
[18] Foucault 1977: 150.
[19] Foucault 1977: 151 (note, notions of subjugation and domination are not quite the same as power – we discuss the features of power later).
[20] Foucault 1977: 154.

chance' and he repeats this idea when he writes: 'the world we know . . . is a profusion of entangled events'.[21] In point of fact, for Foucault, power and contingency conjoin. 'Chance is not simply the drawing of lots, but raising the stakes in every attempt to master chance through the will to power, and giving rise to the risk of an even greater chance'.[22] The result is a history far from that which would comfort us. Foucault asserts: 'We want historians to confirm our belief that the present rests upon profound intentions and immutable necessities. But the true historical sense confirms our existence among countless lost events, without a landmark or point of reference'.[23]

Foucault is faithful, too, to the implications of Nietzsche's historical nominalism. All existent items are historically constructed through the profusion of power and contingency. This holds true for our most precious essentialist notions, like the self, the body, truth and our feelings and very human drives. These are all products of history and bear the scars of their journey through the haphazard power struggles that denote historical formation. For example, 'descent attaches itself to the body', Foucault writes. 'It inscribes itself in the nervous system, in temperament, in the digestive apparatus; it appears in faulty respiration, in improper diets, in the deliberated and prostate body of those whose ancestors committed errors'.[24]

POWER

We have described power as a driving force to history, but it is so central a concept to Foucault's entire project that it demands further scrutiny. Our first question is what power actually is: How does he define the concept? For an answer, Foucault makes clear that power has no definition. This answer does not surprise us, given Foucault's rejection of essentialism and his adoption of Nietzsche's nominalist position. Power, like all items that exist is constituted through history and as such, it cannot be defined. Paraphrasing Nietzsche, 'only that which has no history can be defined'. As an historical item, power changes over time: Foucault writes, 'power is constantly being transformed'.[25] Moreover, power, like all other existent items, is itself historically constructed: 'techniques of power are

---

[21] Foucault 1977: 155.
[22] Foucault 1977: 155.
[23] Foucault 1977: 155.
[24] Foucault 1977: 147.
[25] Foucault 1980: 159.

invented'.[26] The nature of the forms of power, how it operates, all these change according to history. Rather than looking for an essence to power, we need, therefore, to try to describe it.

When we think of describing power in society we often begin by imagining where power resides. Is it in the nation-state, large financial companies, the will of certain individuals or indeed, embedded in the socio-economic structures and practices of our society as Marxists would have us believe? Foucault's analysis of where power resides is rather unusual. He counters three strongly held views. First, against political scientists he argues that power does not always reside in a sovereign power, all be that a personage of a sovereign or the political state – a democratically elected, legal, sovereign body. Foucault writes in an oft-cited passage that is worth quoting at length:

Sovereign, law and prohibition formed a system of representation of power which was extended during the subsequent era by the theories of right: political theory has never ceased to be obsessed with the person of the sovereign. Such theories still continue to day to busy themselves with the problem of sovereignty. What we need, however, is a political philosophy that isn't erected around the problems of sovereignty, not therefore around the problems of law and prohibition. "We need to cut off the King's head: in political theory that has still to be done".[27]

Secondly, against liberals, he counters two points. On the one hand, he opposes the notion that power is somehow something one can possess. He explains:

in the case of the classic, juridical theory, power is taken to be a right, which one is able to possess like a commodity, and which one can in consequence transfer or alienate, either wholly or partially, through a legal act or through some act that establishes a right, such as takes place through cession or contract. Power is that power which every individual holds, and whose partial or total cession enables political power or sovereignty to be established. This theoretical construction is essentially based on the idea that the constitution of political power obeys the model of a legal transaction involving a contractual type of exchange . . . [28]

In short, Foucault says "What we need, however, is a political philosophy that isn't erected around the problems . . . of law and prohibition".[29]

On the other hand, he also argues against what he perceives as the related view, that power is a product of individual wills (be these individual

[26] Foucault 1980: 161.
[27] Foucault 1980: 121.
[28] Foucault 1980: 88.
[29] Foucault 1980: 121.

or collective). Herein, individuals' intentions are often seen as the basis
of the operation of power. He writes:

A second methodological precaution urged that the analysis should not concern
itself with power at the level of conscious intention or decision; that it should not
attempt to consider power from its internal point of view and that it should refrain
from posing the labyrinthine and unanswerable question: 'Who then has power
and what has he in mind? What is the aim of someone who possesses power?'
Instead, it is a case of studying power at the point where its intention, if it has
one, is completely invested in its real and effective practices.[30]

   Finally, he also opposes the Marxist view that power belongs to the
socio-economic system. He poses a number of questions: 'Is power always
in a subordinate position relative to the economy? Is it always in the
service of, and ultimately answerable to, the economy? Is its essential end
and purpose to serve the economy? Is it destined to realise, consolidate,
maintain and reproduce the relations appropriate to the economy and
essential to its functioning? In the second place, is power modelled upon
the commodity'.[31] However, in the end Foucault rejects the nuances of
these questions and argues in favour of a non-economic analysis of power.
He argues that whilst 'power in Western capitalism was denounced by the
Marxists as class domination . . . the mechanics of power in themselves
were never analysed'.[32] Economic-based analyses, such as Marxism, were
always bound to fall short. Power remained poorly understood 'so long as
the posing of the question of power was kept subordinate to the economic
instance and the system of interests which this served'.[33] He, therefore,
asks 'what means are available to us today if we seek to conduct a non-
economic analysis of power?'[34] He concludes, 'very little' and sets out to
develop this analysis himself. His view, in short, is that power does not
reside in a Sovereign state, an individual will or indeed the economic
realm.

   Where then does Foucault perceive power to lie? In fact, his answer
is rather vague. On the one hand, he claims that power has no centre,
no individual single locus. On the other hand, he claims that power is
everywhere. Power, Foucault argues, is dispersed in many different layers

[30] Foucault 1980: 97.
[31] Foucault 1980: 89.
[32] Foucault 1980: 116.
[33] Foucault 1980: 116.
[34] Foucault 1980: 89.

throughout society. If we find the answer to the question of where power resides rather frustrating, let us turn then to the issue of how power works.

Although we cannot define it, we can describe power, the way it operates and how it is dispersed in society. One of the most important elements of power is in fact, the way it operates. It exists in and through relationships. Power is less imposed from an external source, all be that the state or a sovereign. More to the point, power is transmitted through all human relationships, however microscopic.

Power must be analysed as something which circulates, or rather as something which only functions in the form of a chain. It is never localised here or there, never in anybody's hands, never appropriated as a commodity or piece of wealth. Power is employed and exercised through a net-like organisation. And not only do individuals circulate between its threads; they are always in the position of simultaneously undergoing and exercising power. They are not only its inert or consenting target; they are always also the elements of its articulation. In other words, individuals are the vehicles of power, not its points of application.[35]

Given that we are all as individuals enmeshed in and transmitting power, are we all, therefore, conspirators in the fate of our own subjugation? The answer to this question leads us to an assessment of the normative dimension of Foucault's analysis. However, Foucault's normative assessment of power is also seemingly rather opaque. Most social and political analysts use the term 'power' with negative connotations. However, for Foucault the issue of whether power is inherently good or bad is complex. Again, with uncanny resemblance to Nietzsche's views, power is neither negative nor positive. In fact it can be both in two senses.

First, power is neither repressive nor constructive but can be either. In fact, Foucault is keen not to limit a notion of power to a merely repressive one. He writes 'the notion of repression is a more insidious one'.[36] He takes pains to emphasise the constructive aspect of power.

But it seems to me now that the notion of repression is quite inadequate for capturing what is precisely *the productive aspect of power*. In defining the effects of power as repression, one adopts a purely juridical conception of such power, one identifies power with ways which say no, power is taken above all as carrying the force of prohibition. Now I believe that this is a wholly negative, narrow, skeletal conception of power, one which has been curiously widespread. If power

[35] Foucault 1980: 98.
[36] Foucault 1980: 118.

were never anything but repressive, if it never did anything but to say no, do you really think one would be brought to obey it? What makes power hold good, what makes it accepted is simply the fact that it doesn't only weigh on us as a force that says no, but that it traverses and produces things, it induces pleasure, forms knowledge, produces discourse. It needs to be considered as a productive network which runs through the whole social body, much more than as a negative instance whose function is repression.[37]

As Foucault writes, giving the example of sexuality, 'Sexuality' is far more of a positive product of power than power was ever repression of sexuality'.[38]

Secondly, power can be positive or negative in the sense that it may do good or evil. As Foucault notes 'it induces pleasure, forms knowledge, produces discourse'. Not all these items historically constructed through the operation of power are simply instances of evil.

A final point to note about power in Foucault's analysis is a seeming contradiction. On the one hand, he accredits power with an ontological status. He writes: 'It seems to me that power is 'always already there', that one is never 'outside' it . . . '.[39] Herein, power appears to be part of the nature of existence. According to this conception of power, it plays a constituting role, like history itself. Power and history serve to constitute other items. However, how are we supposed to connect such a conception with the idea that power, like all other items, is itself historically consti-tuted? How can power both be constituting and constituted? It seems that Foucault is rather inadequate in addressing this point. The best sense we can make of his position is that power, on the one hand, is constituting and does have some kind of ontological status. It acts alongside history to constitute the ever changing items that compose our world. On the other hand, it is itself constituted in specific forms. It is itself one of the ever-changing items that compose our world. That is to say, specific kinds of power like Christian morality and the human sciences regime, are themselves historically constituted.

In sum, Foucault's depiction of power is that:

1. Power has no essence and as such cannot be defined. We can, however, describe forms of power.
2. Power is historically constituted and its nature changes over time.
3. Power is everywhere but has no locus.

[37] Foucault 1980: 118–19.
[38] Foucault 1980: 120.
[39] Foucault 1980: 140.

4. Power operates through human relationships.
5. Power is neither good nor bad, constructive nor destructive, it can be all these things. A negative instance of power would involve subjugation and domination. A positive instantiation would presumably entail the construction of humanly beneficial items.
6. Power is both historically constituting and constituted.

POWER AND TRUTH

Foucault follows Nietzsche in emphasising the centrality of power in any undertaking to survey the features of society. However, whereas Nietzsche was interested in morality and its genealogy driven by power struggles, Foucault is more interested in truth and knowledge. These two concepts do, of course, display similarities. On the one hand, morality involves moral truth and moral knowledge. On the other hand, forms of truth and knowledge have certain implications for moral beliefs. Foucault's focus is on truth and knowledge and the meshing of these with epistemological and social issues rather than a specific moral emphasis. This shift is no doubt indicative of his times. Christianity's influence was reduced during the twentieth century and the focus upon epistemological and social issues increased proportionally. It is, therefore, to this last that Foucault has turned.

Foucault's genealogy, in looking to truth and knowledge, in point of fact thereby takes on one of our most precious cultural items, truth itself. Foucault writes:

Moreover, the very question of truth, the right it appropriates to refute error and oppose itself to appearance, the manner in which it developed (initially made available to the wise, then withdrawn by men of piety to an unattainable world where it was given a double role of consolation and imperative, finally rejected as a useless notion, superfluous, and contradicted on all sides) – does this not form a history, the history of an error we call truth?[40]

The particular truth that Foucault wishes to analyse is that of his own contemporary times, which, as we have seen, he denoted by the expression 'the human sciences regime'. This regime represents, Foucault believes, the dominant culture of his times. In order to analyse it, Foucault draws together the truth/power nexus that this represents. He makes two points.

[40] Foucault 1977: 144.

First, that truth and power are always interconnected. Foucault explains: "The important thing here, I believe, is that truth isn't outside power, or lacking in power: contrary to a myth whose history and functions would repay further study, truth isn't the reward of free spirits, the child of protracted solitude, nor the privilege of those who have succeeded in liberating themselves. Truth is a thing of this world: it is produced only by virtue of multiple forms of constraint. And it induces regular effects of power. Each society has its regime of truth, its 'general politics' of truth: that is, the types of discourse which it accepts and makes function as true; the mechanisms and instances which enable one to distinguish true and false statements, the means by which each is sanctioned; the techniques and procedures accorded value in the acquisition of truth; the status of those who are charged with saying what counts as true".[41]

Secondly, the nature of this interconnection between truth and power varies throughout history and can only be known genealogically. He, therefore, does a genealogy of the specific truth/power regime of his own times.

We can begin by examining the features of the general relationship between truth and power as envisaged by Foucault. It is poignant to note how distinct Foucault's views about this relationship are from those beliefs dearest to his predecessors of the eighteenth century. His views are in fact in direct opposition to those of the Enlightenment, which purported that truth is something 'objective' and not, 'in essence', connected with power. Indeed, power if it is involved in the process of acquiring truth at all acts only to subvert and distort according to the code of the eighteenth century. Moreover, from an Enlightenment perspective, power is often seen to actually oppose truth: it generates delusional ideologies, rather than objective truths. Power is never perceived to be in the role of actually constructing truth. Finally, truth is something objective, outside of the vagaries of both power and historical change. In short, Enlightenment views equate truth as objective, absolute and free from historical change. Truth is always free from power, and in so far as the latter is involved it can merely serve to distort or repress.

For Foucault, against Enlightenment beliefs truth and power are always intertwined. He writes: 'I believe that the problem does consist in drawing a line between that in a discourse which falls under the category of scientificity or truth, and that which comes under some other category, but

[41] Foucault 1980: 131.

in seeing historically how effects of truth are produced within discourses which in themselves are neither true nor false'.[42]

In turn, each Enlightenment view is challenged. First, Foucault asserts, no truth establishes itself without power. As he expresses it 'Truth is a thing of this world . . . it induces regular effects of power.'[43] He makes much of the particular instance of the figure of truth in whose authority the generation of our knowledge resides, namely the intellectual. The intellectual is, 'the figure in which the functions and prestige . . . are concentrated'. Although old forms of the intellectual are dying, "this new intellectual is no longer the 'writer of genius', but that of the 'absolute savant', no longer he who bears the values of all, opposes the unjust sovereign of his ministers and makes his cry resound beyond the grave". Foucault claims that in spite of the drastic change of the form of the intellectual, he is still bound in power relations. 'Of the intellectual', Foucault writes 'it is he who, along with a handful of others, has at his disposal, whether in the service of the State or against it, powers which can either benefit or irrevocably destroy life'.[44]

Second, no power operates without involving itself in the conception of truth. Foucault explains: 'Each society has its regime of truth, its general politics of truth: that is, the types of discourse which it accepts and makes function as true; the mechanisms and instances which enable one to distinguish true and false statements, the means by which each is sanctioned; the techniques and procedures accorded value in the acquisition of truth; the status of those who are charged with saying what counts as true'.[45]

Third, as we have seen, Foucault believes that power's role is not merely to repress truth. He does not deny that this can occur, but in his view, this is far too narrow a perspective on the relations between power and truth. He writes: 'The notion of repression is a more insidious one, or at all events I myself have had much more trouble in freeing myself from it, in so far as it does indeed appear to correspond so well with a whole range of phenomena which belong among the effects of power'.[46] He repeats this scepticism towards the notion that power only represses the truth when he states 'But it seems to me now that the notion of repression is quite inadequate . . . '.[47] His point is that truth can be repressed by power, but,

[42] Foucault 1980: 118.
[43] Foucault 1980: 131.
[44] Foucault 1980: 129.
[45] Foucault 1980: 131.
[46] Foucault 1980: 118.
[47] Foucault 1980: 119.

importantly, that it can also be constructed by it. Notions of ideology and repression miss this fact. They are 'quite inadequate for capturing what is precisely the productive aspect of power'.[48]

This is an important point in Foucault's analysis, which he takes some lengths to explain.

In defining the effects of power as repression, one adopts a purely juridical conception of such power, one identifies power with a law which says no, power is taken above all as carrying the force of prohibition. Now I believe that this is a wholly negative, narrow, skeletal conception of power, one which has been curiously widespread. If power were never to do anything but to say no, do you really think one would be brought to obey it? What makes power hold good, what makes it accepted, is simply the fact that it doesn't only weigh in on us as a force that says no, but that it traverses and produces things....[49]

In short, Foucault questions 'why the West has insisted for so long on seeing the power it exercises as juridical and negative rather than as technical and positive'.[50]

Fourth, Foucault proposes that power should not always been seen as negative either in the sense of repressive or indeed, in the sense of being evil. Power in relation to truth can be evil. He writes of 'the power of truth' and its attachment to 'the forms of hegemony, social, economic and cultural within which it operates at the present time'.[51] However, power cannot, in his own mind, merely distort truth in favour of evil. He writes 'The essential political problem for the intellectual is not to criticise the ideological contents supposedly linked to science, or to ensure that his own scientific practice is accompanied by a correct ideology...' He continues, 'Its not a matter of emancipating truth from every system of power...

Power can be morally, socially and politically positive. Foucault explains that in fact, the essential problem for the intellectual is 'that of ascertaining the possibility of *constituting* a new politics of truth. The problem', he reiterates, 'is not changing people's consciousness – or what's in their heads – but the political, economic, institutional regime of the *production* of truth'.[52] Just as power can be good or evil, so too, the truths it constructs can be good or evil, and sometimes, perhaps neither.

Fifth, power/truth regimes change though history. They have genealogies.

[48] Foucault 1980: 119.
[49] Foucault 1980: 119.
[50] Foucault 1980: 121.
[51] Foucault 1980: 133.
[52] Foucault 1980: 133.

An issue arising from power's genealogy is that of 'resistance'. We have discussed the genealogical trajectory of items' formations over history. We have seen in the multiple paths of 'development' both ruptures and discontinuities. What Foucault really emphasises, however, is that these ruptures do not mean that the old regimes of truth and power are simply obliterated whilst the new ones emerge. He asserts that old truth-power regimes remain and form resistance to the new ones imposed upon the old. He explains 'that there are no relations of power without resistances'.[53] In fact, he continues: 'the latter [relations of resistance] are all the more real and effective because they are formed right at the point where relations of power are exercised; resistance to power does not have to come from elsewhere to be real, nor is it inexorably frustrated through being the compatriot of power. It exists all the more by being in the same place as power; hence, like power, resistance is multiple and can be integrated in global strategies'.[54]

Finally, there are very specific ways in which truth and power are related. Foremost among these is the idea that power is internal to truth. It does not merely impose itself from above, like a grand sovereign or external body of rules. Instead, it acts from within truth – the productions and relationships of truth. Foucault sums up these sentiments when he writes: "At this level it's not so much a matter of knowing what *external* power imposes itself on [truth], as to what effects of power circulate among [truths], what constitutes, as it were, their *internal* regime of power . . . [55]

THE HUMAN SCIENCES REGIME

Having witnessed the features Foucault attributes to the relationship between truth and power generally throughout history, we might now move on to witness those which he ascribes to the very specific truth/power regime of his contemporary times. This is the historical context of the liberal era of the 1960s which, according to Foucault, saw the growth and consolidation of what he termed the 'regime of the human sciences'. Whereas Christianity had been the predominant culture of Nietzsche's times, the human sciences regime constituted the new truth-power regime of Foucault's era.

For a definition of what Foucault means by the human sciences regime, we encounter the difficulties of his nominalist approach. This regime is

[53] Foucault 1980: 142.
[54] Foucault 1980: 142.
[55] Foucault 1980: 113.

described rather than strictly defined. However, we can see that Foucault's notion of a regime encompasses, firstly, a set of discourses. He refers to these as the 'discourse of the human sciences'.[56] However, these discourses are not synonymous with a precise discipline. He explains:

But neither do they coincide with what we ordinarily call a science or a discipline even if their boundaries provisionally coincide on certain occasions; it is usually the case that a discursive practice assembles a number of diverse disciplines or sciences or that it crosses a certain number among them and regroups many of their individual characteristics into a new and occasionally unexpected unity.[57]

Secondly, the human sciences regime includes a range of institutions; thirdly, a body of practices; and finally a complex web of relationships. Foucault captures these points when he writes:

Discursive practices are not purely and simply ways of producing discourse. They are embodied in technical processes, in institutions, in patterns for general behaviour, in forms for transmission and diffusion, and in pedagogical forms which, at once, impose and maintain them.[58]

The human sciences regime refers to all the institutions of the social sciences and in particular those with a therapeutic or educational role. It includes the languages, practices and relations of hospitals, asylums, prisons, schools and other corrective institutions.

The Human Sciences regime is very much Foucault's view of our contemporary cultural counterpart to Nietzsche's Christian 'regime': the human sciences are to modern societies what Christianity was to earlier centuries. They are discourses that present themselves as truths. Just as Christianity hid its historical constitution and presented its truths as absolute, so, too, according to Foucault does the human sciences regime. Whilst Christian truths gained their absolute status through religious criteria, human sciences' truths gain theirs through reference to empirical criteria. These truths are presented as objective 'discoveries'. However, as genealogies demonstrate, the truths of the human sciences regime are in fact constructed through a historically emergent truth/power regime.

This truth-power regime is historically constituted through a power struggle. The human sciences regime is in competition with old power regimes like Christianity. It needs to impose its new truths upon the old truths. For example, just as in Enlightenment the natural sciences

---

[56] Foucault 1980: 107.
[57] Foucault 1980: 200.
[58] Foucault 1980: 200.

competed with Christianity and produced a new truth about the origins of the world and the nature of the universe and indeed mankind, the power regime of the human sciences needs to impose its supposedly detached scientific knowledge about the norms of behaviour and the essence of human nature to construct new standards for social behaviour. "These are not simply new discoveries, there is a whole regime in discourse and forms of knowledge".[59] The human sciences whilst purporting to show us an a-historical truth about the essence of human beings and correct forms of behaviour, in fact can be demonstrated through genealogy to be a historically emergent regime, with multiple lines of descent, formed through power and contingency. There are certain ironies here pointed out in *Power/Knowledge*. For instance, the human sciences regime often depicts itself as anti-Establishment and outside of mainstream sources of power and authority. This discourse, itself, however is simply part and parcel of its own truth-power regime.

Foucault is at pains to point out how the regime of the human sciences generates a self-perception of itself as neutral, objective and outside the political arena and consequently any relations of power. However, for Foucault, power is intrinsic to all forms of knowledge. Truth and knowledge in the human sciences are constituted by and enmeshed in power.

Furthermore, the human beings and relations surrounding these discourses of truth are also caught up in power. Although doctors, social workers, psychiatrists, educational therapists etc. believe themselves part of a benign and politically neutral regime, Foucault, argues that all human relationships transmit power, including those of the human sciences. Consequently, all these practitioners are caught up in a web of power. They transmit power, help construct truths and generate a hierarchy of truth. If you were to see a psychotherapist, for example, power is transmitted in that very relationship. Forms of knowledge are being conveyed which subject you to a particular discourse and through the power regime those truths are implanted into you and generate new beliefs and new forms of behaviour.

A further important point to note about the human sciences is its perception of power as 'bad' and itself as 'good'. It perceives itself as concerned with the welfare of the human population, its forms of knowledge serve only the human good, its institutions are here not to disempower, control or exploit us, but to help and enable us. Indeed, those

[59] Foucault 1980: 12.

working in the human sciences evoke the common perception of power as a hostile force and themselves as completely independent of it.

Finally, power relations in the human sciences regime do not arise simply from the political state or economic companies controlling science. But the power operates within science itself, its forms of knowledge and own internal relations. Foucault sums up these sentiments when he writes: "At this level it's not so much a matter of knowing what external power imposes itself on science, as to what effects of power circulate among scientific statements, what constitutes, as it were, their internal regime of power, and how and why at certain moments that regime undergoes a global modification".[60]

DISCIPLINARY REGIME OF THE HUMAN SCIENCES

The Human Sciences Regime has distinct features by which power operates and truth is constructed. Foucault describes the most important features of these in a passage worth quoting at length. He writes:

... along with all the fundamental and technical inventions and discoveries of the seventeenth and eighteenth centuries, a new technology of the exercise of power also emerged which was probably even more important than the constitutional reforms and new forms of government established at the end of the eighteenth century. In the camp on the Left, one often hears people saying that power is that which abstracts, which negates the body, represses, suppresses, and so forth. I would say that what I find most striking about these new technologies of power introduced since the seventeenth and eighteenth centuries is their concrete and precise character, their grasp of a multiple and differentiated reality. In feudal societies power functioned essentially through signs and levies. Signs of loyalty to the feudal lords, rituals, ceremonies and so forth, and levies in the form of taxes, pillage, hunting, war etc. In the seventeenth and eighteenth centuries a form of power comes into being that begins to exercise itself through social production and social service. It becomes a matter of obtaining productive service from individuals in their concrete lives. And in consequence, a real and effective incorporation of power was necessary, in the sense that power had to be able to gain access to the bodies of individuals, to their acts, attitudes and modes of everyday behaviour. Hence the significance of methods like school discipline, which succeeded in making children's bodies the object of highly complex systems of manipulation and conditioning. But at the same time, these new technologies of power needed to grapple with the phenomena of population, in short to undertake the administration, control and direction of the accumulation of

---

[60] Foucault 1980: 113.

men . . . Hence there arise the problems of demography, public health, hygiene, housing conditions, longevity and fertility.[61]

Let us now analyse the individual features of what Foucault describes as the modern 'disciplinary regime'. Foucault refers to the human sciences power/truth regime as disciplinary because it disciplines the human body and mind. He writes "Discipline makes individuals, it is the specific technique of a power that regards individuals both as objects and as instruments of its exercise."[62] That is to say, it constructs along the lines of the truths that it imposes upon us. For example, it creates notions of normal (e.g., sane or socially acceptable) behaviour. It then has sciences to elaborate these truths and institutions to construct these forms of behaviour.

Foucault states that 'The success of disciplinary power derives no doubt from the use of simple instruments; hierarchical observation, normalising judgement and their combination in a procedure that is specific to it, the examination'.[63] He gives details of the ways in which the institutions of the human sciences achieve this specific discipline of us.

There are five main ways in which Foucault describes the disciplinary regime. First, it operates through *surveillance*. It *investigates, examines, records* and *diagnoses* our behaviour. For example, when we attend a doctor's surgery, we have our symptoms described and recorded, our bodies subjected to external and internal examination. Detailed records are kept on us and diagnosis of our abnormality then made. Consider a second example, the school. Just as our bodies are examined in the doctor's surgery, in schools, we sit exams and have our mind's examined. Results are then recorded from the examination and we are then 'diagnosed' as passing or failing, etc.

Secondly, the modern power/truth regime of the human sciences, according to Foucault, *categorises* us. It offers categories of types of afflictions. These categories are an ever-increasing list of 'syndromes', 'afflictions', 'abnormalities'. For example, there is no longer a criminal who is an ordinary human being who has transgressed morally. Instead, there is a type, a psychotic, a socio-path or a schizophrenic. Likewise, in the case of sexual categorisation, there is no longer a man or woman who performs the act of sleeping with the same sex, there is a *category* of the homosexual. In the school examination hall, there is not somebody who

[61] Foucault 1980: 124.
[62] Foucault 1979: 170.
[63] Foucault 1979: 170.

merely cannot read or spell in the way the 'normal' majority do, there is the *type* of the dyslexic.

Thirdly, the modern disciplinary regime *normalises* us. Foucault writes that "the power of the Norm appears through the disciplines".[64] He adds that "Like surveillance and with it, normalisation becomes one of the great instruments of power at the end of the classical age".[65] The human sciences create a notion of what is normal human nature, normal human behaviour, all be that mental, physical, sexual, social. It creates a notion of what is normal in all aspects of our lives and then measures all cases of human behaviour against this. It seeks to normalise by analysing and creating categories of abnormality, for example, schizophrenia, neurosis, criminal behaviour and bodily disorders. If we have any of these, or indeed other afflictions, we can go to a specialist in the human sciences and be normalised in one of its institutions.

Fourthly, the institution of the human sciences normalises us by *training* us: 'the chief function of the disciplinary power is to "train"'.[66] Through training, we are constructed as normal individuals. This training has a number of features.

i. Foucault explains that "the power of normalisation . . . individualizes by making it possible to measure gaps".[67] These 'gaps' can be those of time or space. Our first feature of training is in fact that it occurs through the regimentation of *time*. Time is divided and regulated. For example, the duration of educational examinations, the length of classes at school, the timetable of the clinic, the waiting rooms at the doctor or dentist serve to help regulate that time.

ii. *Space* is regulated. Rooms are created for special use and then divided off. For example, we have the space of offices for work; prison cells for penal correction; examination halls for mental assessment; classrooms for learning; hospital wards for the correction and normalisation of the sick body.

iii. Training occurs through a strict hierarchy. It "hierarchizes qualities, skills and aptitudes".[68] In the human sciences regime, there are experts and non-experts. For example, there is the social worker – the expert on assessing the norms of social behaviour. First in command is this expert and the hierarchy of experts from the senior, most qualified

[64] Foucault 1979: 184.
[65] Foucault 1979: 184.
[66] Foucault 1979: 170.
[67] Foucault 1979: 184.
[68] Foucault 1979: 181.

through to the most junior-qualified person. Second, there is the trainee and the hierarchy of trainees according to age and experience etc. Finally, there is the unqualified layperson. At the bottom of the hierarchy, however, is the patient, or criminal. The person to be assessed and normalised. Foucault sums up, 'The perpetual penalty that traverses all points and supervises every instant in the disciplinary institutions compares, differentiates, *hierarchizes*, homogenises, excludes. In short, it normalizes'.[69]

iv. Training also occurs through *drills*. These are practices of repetitive behaviour for installing norms. Foucault describes the drill of prisoners being exercised in the yard. He offers the example of children at school being given the drill of the school assembly.

v. Finally, the disciplinary regime works by a form of 'internalisation': you operate the power regime upon yourself. You categorise yourself. For example, you perceive yourself to have had a normal or disturbed childhood; a normal body or one that is too fat or too thin. You eat correctly or you have eating disorders – bulimia or anorexia. You are the right shape or the wrong shape. You have good posture, movement and fitness, otherwise you diagnose yourself as incorrect; you undergo your own training, perhaps in the gym, to normalise and correct your body. You thereby train *yourself*, construct your own normality.

Foucault summarises these points in the following quotation: 'Truth is to be understood as a system of ordered procedures for the production, regulation, distribution, circulation and operation of statements . . . Truth is linked in a circular relation with systems of power which produce and sustain it, and to effects of power which it induces and to which extend it. A regime of truth'.[70]

The traits of the modern power regime are that, firstly, it surveys. This is to say that it investigates, examines, records and diagnoses. Secondly, categorises types of afflictions. 'abnormalities'. E.g., no longer a criminal but a type – psychotic, schizophrenic. Thirdly, the modern disciplinary regime normalises – it creates a notion of what is normal and measures all other categories against this. Fourthly, it trains. The modern disciplinary regime constructs normal individuals through the division of time, space, the institutionalisation of hierarchy and the installation of drills. Finally, you operate power from the disciplinary regime on yourself.

---

[69] Foucault 1979: 183 (my emphasis).
[70] Foucault 1980: 133.

We can display a simplified version of these features of the modern disciplinary truth/power regime of the human sciences in the following chart.

DISCIPLINE

1. Surveillance. It investigates, examines, records and diagnoses our behaviour.
2. Categorisation. Types of afflictions. Increasing list of 'syndromes', 'afflictions', 'abnormalities'.
3. Normalises. Notion of what is normal, measures all other categories against this.
4. Trains. Constructs normal individuals.
   a. Time
   b. Space
   c. Hierarchy
   d. Drills
5. 'Self-discipline'. Internalisation of power regime.

In his conception of the human sciences regime Foucault has adopted Nietzsche's notion of the historical construction of society through power and contingency. He has also borrowed Nietzsche's approach of genealogy to understand the construction of knowledge in modern societies and the relationship of truth to power. Foucault expresses these concerns and his debt to Nietzsche when he writes, 'The political question, to sum up, is not error, illusion, alienated consciousness or ideology; it is truth itself. Hence the importance of Nietzsche".[71]

---

[71] Foucault 1980: 133.

8

Applications of Genealogy

We have studied Foucault's philosophical perspectives on the human sciences. His approach centres around concepts of Power/Knowledge and the theory of genealogy. Our final purpose is to look at how he deploys these philosophical approaches to practical, dare I say, 'empirical' instances in the human sciences. Here we can do no better than continue with Foucault's own work and examples of genealogy. First, we conduct a detailed study of his genealogy of the penal system. Secondly, we examine his work on the history of sexuality.

PENAL INSTITUTIONS

For our example of the application of Foucault's approach of genealogy we look first to the particular institutions and practices of criminal reform. Foucault studies these in his seminal *Discipline and Punish*. Foucault analyses our modern disciplinary regime by doing a genealogy. He looks to the heterogeneous past of the penal regime – its transitions through the historical power struggles by which it is constituted. There are three main stages of historical change and rupture that result in the constitution of the contemporary penal system, according to Foucault. These are firstly, torture, secondly, punishment and finally, modern discipline.

TORTURE

Let us look first at the practice of torture. Foucault begins his genealogy of the penal system in true characteristic style, with a flamboyant attempt to shock. It is not for the faint-hearted. Foucault writes "On 2 March 1757

Damiens, the regicide, was condemned before the main door of the Church of Paris, where he was to be taken and conveyed in a cart, wearing nothing but a shirt, holding a torch of burning wax weighing two pounds; then, 'in the said cart, to the place de Greve, where on a scaffold that will be erected there, the flesh will be torn from his breasts, arms, thighs and calves with red-hot pincers, his right hand, holding the knife with which he committed the said parricide, burnt with sulphur, and, on those places where the flesh will be torn away, poured molten lead, boiling oil, brining resin and sulphur melted together and then his body drawn and quartered by four horses and his limbs and body consumed by fire, reduced to ashes and his ashes thrown to the winds".[1] This is Foucault's depiction of the horrors of the penal regime that occurred before our own.

Foucault goes on to analyse the specific features of torture. He depicts four aspects to the penal system, which can be discussed under the notions of visibility, truth, punishment and revenge.

Visibility is very important and is of a very particular nature in the torture regime. It occurs in the form of the *Spectacle*. A crowd watches an elaborate and horrific death of the condemned. The signs of punishment are visibly inflicted upon the human body. These wounds and mutilations upon the body are the visible signs of an almighty and terrible power. This power is therefore highly visible. The spectacle of torture has the aspect of the carnival. The rules are inverted, and indeed authority, too. The mocked criminal is turned into a hero and there is evidence of the vengeance of the people against the monarch. There is always a danger that this inversion could go too far and the sovereign's power be usurped.

The second element if that of truth. Torture is carried out to achieve a confession. This confession then acts as proof of the crime. The truth is extracted through the torture, bit by bit. Thirdly, the regime of torture includes the phenomenon of punishment. As each piece of evidence is extracted though torture, each piece of truth constructed, so, too, is a piecemeal punishment dealt out.

Finally, torture, according to Foucault, demonstrates revenge. This is the revenge of the all powerful sovereign.

In sum, torture consist in the highly visible extraction of confession – hence truth – punishment and almighty, terrible revenge.

---

[1] Foucault 1979: 3.

PUNISHMENT

The next layer in history Foucault refers to as the punishment regime. This is punishment without revenge. It is a more humane form of punishment which is less horrific, for example, the electric chair. This regime retains elements of the previous torture era as the historical layers of the various power struggles accumulate. Foucault pays scant attention to punishment in comparison with torture and discipline which are the mainstay of his analysis. Let us, therefore, move on to focus upon Foucault's own preoccupations.

DISCIPLINE

The modern penal regime belongs to the layer of history characterised by the modern disciplinary regime. Foucault analyses the modern penal system through the same key notions of visibility, truth, punishment and revenge. It has the following features.

First, discipline, too, has a very specific form of visibility. A perfect example of this is Bentham's panoptican. Foucault writes of the kind of architecture of which the panoptican is archetypal: "that of an architecture that is no longer built simply to be seen or to observe the external space but to permit an internal, articulated and detailed control – to render visible those who are inside it; in more general terms, an architecture that would operate to transform individuals: to act on those it shelters, to provide a hold on their conduct, to carry the effects of power right to them, to make it possible to know them, to alter them".[2]

In the panoptican, individual prisoners are all highly visible in their cells. A *few* prison guards can thereby watch *many* prisoners from their observing tower: The perfect disciplinary apparatus would make it possible for a single gaze to see everything constantly.[3] Note the contrast here to the torture regime. In this spectacle, many watched the few. In the disciplinary regime, a few watch the many. Moreover, the few are invisible to the many. In fact, it is even more 'secretive' than that. The architectural design of the panoptican is such that you don't need anyone to watch at all. You can't see if anyone is watching you in the tower, so you feel as if you are being watched all the time, even if the tower is empty! The surveillance is embedded within the very architecture itself. The

[2] Foucault 1979: 172.
[3] Foucault 1979: 173.

result of this is that individual's respond as if they are being watched. The result, in turn, of this, is that prisoners undertake self-surveillance. They monitor themselves. Furthermore, even the prison wardens feel surveyed: Foucault claims that disciplinary power is embedded within the architecture of the institution such that all are caught in an invisible web of power.

This power does not torture for confession or truth. It instigates self-surveillance, self formation and self construction in line with the disciplinary regime. It normalises: the prisoners normalise themselves. It is part and parcel of the same principle as the human sciences, part of their disciplines and practices.

The second feature of the modern penal institution, according to Foucault, is its instantiation of truth. In contrast to the confessions of 'truth' exacted under punishment, this regime enacts a kind of conceptual objectification of intervals. Science investigates the criminals, it produces diagnoses of their personalities, minds, brains and bodies. It then categorises various types of criminals with various abnormalities. Moreover, the regime evokes individuals to examine, diagnose and correct themselves. In this way they 'internalise' and carry forth the truths of the disciplinary power regime of the human sciences.

Finally, against punishment or revenge, the disciplinary regime normalises. Foucault describes how the modern disciplinary regime is not interested in whether you are guilty but rather in why you are guilty. Why did you behave in a criminal way? In contrast to previous regimes, the search is not for the act of the crime, but for the type of the criminal. The main investigation is not centred around what you did, but is a diagnosis of why you did what you did. The diagnosis looks to see what kind of mental, emotional, neurological abnormality or personality defect you are afflicted with. The result is not torture or punishment but treatment and *normalisation*.

Normalisation occurs through the methods previously traced within the disciplinary regime. First, there is training. This occurs through the division and regulation of time, space and social hierarchy. For example, space is divided into cells, time is regulated through the timetable. There is a strong hierarchy of inmates and wardens.

Training also occurs through drills. The body is thereby trained too. Penal systems construct the emotions and generate a normalised, 'healthy mind'. Foucault describes how modern penal systems also construct the disciplined body.

Foucault describes the disciplined body thus: "the body [is] object and target of power. It is easy enough to find signs of the attention then paid to the body – to the body that is manipulated, shaped, trained, which obeys, responds, becomes skilful and increases its forces. The great book of man-the-Machine was written simultaneously on two registers: the anatomico-metaphysical register, of which Descartes wrote the first pages and which the physicians and philosophers continued, and the techno-political register, which was constituted by a whole set of regulations and by empirical and calculated methods relating to the army, the school and the hospital, for controlling or correcting operations of the body".[4]

Through Foucault's genealogical perspective there is no notion that the disciplinary regime developed in a linear or organic fashion through history. The old layers and power regimes were not, therefore, replaced, but remained and formed a constellation of powers. Traces of the old penal regimes of torture and punishment co-existed alongside the new disciplinary regime. The physical treatments and the confinement in cells represented the enactment of multiple roles. They were normalisations and punishments. In any genealogical approach – there is an acceptance of the resistance of the previous power regimes.

HISTORY OF SEXUALITY

'We are not shaped through external forces acting upon us by a policing state or body of external rules; we shape our own behaviour through acting within the modern disciplinary regime: we collude with it and construct ourselves according to its norms'. If this view is made clear in Foucault's depiction of the penal regime, is it perhaps made even more so in the way he argues we construct our deepest and most secret truths, in the realm of our very sexuality. It is to Foucault's *History of Sexuality* that we turn to see a further instance of the application of genealogy.

Foucault depicts immense detail in his Three Part study on the *History of Sexuality*. This enormous body of work embraces issues as diverse as Ancient, Eastern and contemporary Western comparisons, issues of economics, health, identity, morality, pleasure, emotion and family. His main genealogical thesis, however, is developed in Part One, *The History of Sexuality: An Introduction*. Herein he locates sexuality as part of the human sciences power/knowledge regime. Foucault defines his project: 'The Object, in short, is to define the regime of power-knowledge-pleasure that sustains the discourse on human sexuality in our part of the world'.[5]

[4] Foucault 1979: 136.
[5] Foucault 1978: 11.

The bedrock of Foucault's genealogy of modern sexuality can be captured roughly as follows: 'For a long time, the story goes, we supported
a Victorian regime, and we continue to be dominated by it even today.
Thus, the image of the imperial prude is emblazoned on our restrained,
mute and hypocritical sexuality'.[6] He charges, in his inevitable sixties
mode that:

silence became the rule. The legitimate and procreative couple imposed itself
as model, enforced the norm, safeguarded the truth, and reserved the right to
speak while retaining the principle of secrecy. A single locus of sexuality was
acknowledged in social space as well as at the heart of every household, but it was
a utilitarian and fertile one: the parents' bedroom.[7]

This 'classical' image of repressed sexuality is one that he believes is
fairly accepted as a truism. We have inherited a form of sexuality that predominantly hides and represses. This viewpoint he terms the 'repressive
hypothesis'.[8]

However, in its straightforward form, Foucault challenges this conception. He queries; 'Is sexual repression truly an established historical
fact?'.[9] Furthermore, he asks, 'Do the workings of power, and in particular those mechanisms that are brought into play in societies such as ours,
really belong primarily to the category of repression?'.[10]

He answers these questions by pointing out that he does not deny
certain features of the 'repressive hypothesis'. He states: 'Let there be
no misunderstanding: I do not claim that sex has not been prohibited
or barred or masked or misapprehended since the classical age; nor do
I even asset that it has suffered these things any less from that period
than before. I do not maintain that the prohibition into the basic and
constitutive element from which one would be able to write the history
of what has been said concerning sex starting from the modern epoch.'
His point is to argue that the repressive hypothesis is not the full story. He
explains: 'All these negative elements – defences, censorships, denials –
which the repressive hypothesis groups together in one great central
mechanism destined to say no, are doubtless only *component parts* that
have a local and tactical role to play in a transformation into discourse,

[6] Foucault 1978: 1.
[7] Foucault 1978: 3.
[8] Foucault 1978: 10.
[9] Foucault 1978: 10.
[10] Foucault 1978: 10.

a technology of power, and a will to knowledge that are far from being reducible to the former'.[11]

Foucault's belief, following his genealogical principles outlined earlier, is that power, rather than merely exhibiting a prohibitive role, plays an all important constitutive one. He explains, 'In actual fact, what was involved, rather, was *the very production of sexuality*. Sexuality must not be thought of as a kind of natural given which power tries to hold in check, or as an obscure domain which knowledge gradually tries to uncover. It is the name that can be given to a historical construct'.[12]

The repressive hypothesis had, Foucault argued, far from silencing sexuality, produced a particular discourse about it. He writes: 'the central issue, then is not to determine whether one says yes or no to sex whether one formulates prohibitions or permissions not to account for the fact that it is spoken about, but to discover who does the speaking, the positions and viewpoints from which they speak, the institutions which prompt people to speak about it and which store and distribute the things that are said. What is at issue, briefly, is the over-all "discursive fact," the way in which sex is "put into discourse"'.[13]

In order to see the nature of this modern discourse about sexuality, Foucault depicts a genealogy in which nineteenth- (and later twentieth-) century sexuality displaced earlier Christian forms. In his essay on *The History of Sexuality* in *Power/Knowledge*[14] Foucault describes how the power regime of Christianity in fact prefigured certain features of the sexuality of the modern human sciences regime. He describes first, how Christianity on the one hand, authorised certain kinds of sexual activity whilst at the same time prohibiting and punishing others: 'in discussions of the penitential it is always emphasised that Christianity imposes sanctions on sexuality, that it authorises certain forms of it and punishes the rest'.[15] Secondly, he notes a more important productive feature of Christianity, namely the confessional. Whilst the prohibition of sexuality within Christian dogma is often focussed upon, this merely negative power has a much more pervasive positive counterpart. The confessional acted as an early form of surveillance of sexuality and gave rise, that is produced or historically constructed, a whole body of theory and practice about sexuality. Foucault explains: 'But one ought, I think, also to point out

[11] Foucault 1978: 12 (my emphasis).
[12] Foucault 1978: 105 (my emphasis).
[13] Foucault 1978: 11.
[14] Foucault 1980.
[15] Foucault 1980: 186.

that at the heart of Christian penitence there is the confessional, and so admission of guilt, the examination of conscience, and arising from that the production of a whole body of knowledge and a discourse on sex which engendered a range of effects on both theory and practice.'[16]

Nineteenth- and twentieth-century sexuality then emerged and generated a discourse which placed sex at the centre of the mind. Freudian theory of course, is the central instance. Herein sexuality underwent a medicalisation: mental health and illness were seen to be tied up with sexuality.

Psychoanalysis was the discipline in the human sciences regime at the centre of this discourse about sexuality. In psychoanalysis the treatment of mental health occurred in good part through the normalisation of sexuality. He states: 'One can say certainly that psychoanalysis grew out of that formidable development and institutionalisation of confessional procedures which has been so characteristic of our civilisation. Viewed over a shorter span of time, it forms part of that medicalisation of sexuality which is another strange phenomenon of the West'.[17]

Foucault believes that the power/knowledge regime of sex displays the same features as other regimes of the human sciences, namely, it is 'disciplinary'. Its features are characteristic of the modern disciplinary regime of the human sciences.

First, the modern disciplinary regime has the feature of *surveillance*. It investigates, examines, records and diagnoses our sexual behaviour. Foucault depicts an example of this surveillance with respect to child sexuality. He writes: 'there occurred a 'pedagogization of children's sex: a double assertion that practically all children indulge or are prone to indulge in sexual activity; and that, being unwarranted, at the same time "natural" and "contrary to nature", this sexual activity posed physical and moral, individual and collective danger; children were defined as "preliminary" sexual beings, on this side of sex, yet within it, astride a dangerous dividing line. Parents, families, educators, doctors, and eventually psychologists would have to take charge, in a continuous way, of this precious, and perilous, dangerous and endangered sexual potential'.[18]

Secondly, the disciplinary regime *categorises* sexuality. Foucault denotes four categories. He writes 'Four figures emerged from this preoccupation with sex four privileged objects of knowledge, which were also targets and

[16] Foucault 1980: 186.
[17] Foucault 1980: 191.
[18] Foucault 1978: 104 (my emphasis).

anchorage points of the ventures of knowledge: the hysterical woman, the masturbating child, the Malthusian couple, and the perverse adult.'.[19] Moreover, the sexual disciplinary regime denotes types of afflictions, indeed an increasing list of 'syndromes', 'afflictions' and 'abnormalities'. Foucault expresses it thus: 'the sexual instinct was isolated as a separate biological and psychical instinct; a clinical analysis was made of all the forms of anomalies by which it could be afflicted'.[20]

Thirdly, the disciplinary regime *normalises*. It generates a notion of what is sexually normal and measures all other categories against this. He explains that the sexual instinct 'was assigned a role of normalisation or pathologisation with respect to all behaviour.[21]

Finally, the disciplinary regime *trains*. It constructs normal sexual individuals. He explains, with respect to abnormal sexualities, that 'finally a corrective technology was sought for these anomalies'.[22] He also depicts the everyday training of our sexualities. He describes a 'socialisation of procreative behaviour brought to bare on the fertility of procreative couples; a political socialisation achieved through the "responsibilization" of couples with regard to the social body as a whole and a medical socialisation carried out by attributing a pathogenic value – for the individual and the species – to birth-control practices'.[23]

Modern sexuality has a genealogy like any other historical item: it is historically constructed through haphazard pathways and relations of power, from classical and Christian forms of discourse to their modern counterparts. Furthermore, it displays the typical feature of the modern disciplinary truth-power regime, namely, surveillance, categorisation, normalisation and training.

In sum, we have seen an example of the application of this particular historical approach to the humanities, known as genealogy. This approach serves to de-legitimise institutions, that present themselves with a pedigree in order to legitimate their supremacy and hide their origins in power struggles and contingency. We saw earlier how this tradition began with Nietzsche and was taken up and adapted by Foucault. Here, we have witnessed Foucault's application of genealogy to a study of penal regimes and to sexuality. Foucault has shown the layers that make up a regime of 'discipline' and 'normalisation' and contrasted this with earlier

[19] Foucault 1978: 105.
[20] Foucault 1978: 105.
[21] Foucault 1978: 105.
[22] Foucault 1978: 105.
[23] Foucault 1978: 105.

penal regimes based upon torture and punishment and sexual regimes based upon repression and confession. Foucault also shows how some of the features of earlier power regimes remain in our current, supposedly more progressive institutions.

CONCLUSION

We have seen the tradition of continental philosophy of social science known as genealogy. This is a particular historical way of analysing items in the humanities. We have looked at a genealogy as developed by Nietzsche. We have also seen something of how he applied it to the analysis of ethics, the ethics of Christianity. We have seen how genealogy is a form of historical analysis that unmasks as it traces the past; a form of historical analysis that breaks the delusions and pretensions of ideologies, practices and all forms of cultural, social and political institutions. We also have elaborated further details of the tradition of genealogy by seeing how Foucault developed Nietzsche's ideas and used them for the analysis of social institutions, practices and forms of truth/knowledge and social science in our contemporary society.

PART III

CRITICAL THEORY

O ur final tradition is the very broad and highly acclaimed one of Critical Theory. This term is often used very loosely to include thinkers like Foucault, the post-structuralists or those influenced by Nietzsche, Marx and Freud, and indeed some even go so far as to apply the term to almost any critical approach to the humanities at all. It should be kept in mind however that in correct usage *Critical Theory* denotes a very specific tradition.

Critical Theory, properly conceived, is the brainchild of the Early Frankfurt School. It is worth noting the enormous impact of the Early Frankfurt School and critical theory in post war liberal Europe and indeed throughout the world. Held explains that: 'The writings of what one may loosely refer to as a "school" of western Marxism – critical theory – caught the imagination of students and intellectuals in the 1960's and early 1970's. In Germany thousands of copies of the "school"s' work were sold, frequently in cheap pirate editions. Members of the New Left in other European countries as well as in North America were often inspired by the same sources. In other parts of the world, for example in Allende's Chile, the influence of these texts could also be detected. In the streets of Santiago Marcuse's name often took a place alongside Marx and Mao in the political slogans of the day. Critical theory became a key element in the formation and self-understanding of the New Left'.[1]

The Early Frankfurt School was composed of individuals who, in the main, were from cultivated Jewish German families, born in the first half of the twentieth century. They lived their early years against the backdrop

---

[1] Held 1980: 13.

of World War I. These scholars went on to be educated during the time of the moderate 'socialism' of the freshly created Weimar Republic. They all witnessed the build-up of National Socialism and saw Hitler's rise to power. This was the Germany of the 1930s. With their humane sensitivity, the Early Frankfurt School members were appalled by the political events around them and sought refuge in what they perceived as an institution of like-minded colleagues. They set up their institution in the 1930s – formally known as the Frankfurt Institute for Social Research. It was a left-leaning institute for Social Research with an interdisciplinary approach to the study of society and the humanities. The School was in fact, originally set up as an institute for Marxism by Felix Weil, the son of a millionaire. It had been formally opened in 1923 and was first directed by the professorial Marxist Carl Grunberg. Later, directorship was handed on to Max Horkheimer who became Theodor Adorno's great friend and who was later to become co-author of certain seminal texts.[2]

The Institute's principal members besides Horkheimer and Adorno included Eric Fromm, Friedrick Pollock, Leo Lowenthal, Herbert Marcuse, Franz Neumann, Otto Kirchheimer and, of course, more complexly, Walter Benjamin. All were influenced in one way or another, and in varying degrees of depth by forms of Marxism. Their disciplinary orientations centred around the social sciences and spread into areas as diverse as politics, literary theory, art, philosophy, history and psychoanalysis. They spanned from the empirical to the highly abstract and theoretical.

Due to the political barbarism of Germany in the 1930s and 1940s the Institute was forced to flee. Many members went to the New World and Adorno moved with them. Here in North America they lived, researched, thought and wrote until Nazism was defeated in Europe. Then in the mid-1940s the Institute returned to Germany.

Due to their own historical era, critical analysis became the preoccupation of the entire school. In particular, they wanted to know how and why Nazism had arisen in a developed, supposedly civilised society. The Early School members were all animated by this question. They were all involved in a *critical* analysis of their contemporary society from an empirical or theoretical standpoint.

This group of predominantly Jewish thinkers were inspired by the deepest traditions of German philosophy. They looked to the German inheritance of ideas from the eighteenth century on. Particularly, they were inspired by Karl Marx and to a certain extent by Freud. Their

---

[2] See Wiggershauss 1994 for an excellent biography of the Frankfurt School.

critical approach to social and cultural understanding was given form and discipline in the early articulations of Horkheimer. He coined the notion 'critical theory' with a very special approach to theorising. Later Horkheimer's ideas were developed in several directions by contemporaries in the Early Frankfurt School and critical theory was also taken up by later members. One should not underestimate the impact and influence of this tradition and its ubiquitous presence in all walks of the humanities from the mid-twentieth century until the present day.

If hermeneutics can be encapsulated as dealing with interpretation, genealogy as providing a form of historical analysis, then critical theory, as its name suggests, offers us a form of critical understanding. This critical understanding can be applied to many different aspects of Western culture. First, it is a potent form of analysis of culture understood in its broadest sense. Critical Theory is most important in the cultural realm of knowledge. It is highly pertinent to any study of academic disciplines – science, the incursion of natural science into the human sciences, positivism, forms of historical analysis, philosophical speculation – education and of course, ideologies. Secondly, critical theory can also be used in analysis of the features of our society, of our institutions like those concerned with our welfare, hospitals, psychiatric institutions etc. It is more often deployed upon an examination of political institutions and practices, so for instance, parliament, and finally systems of democracy, ideas about freedom; commerce. Third, critical theory concerns itself with culture, in the narrower sense, that is with fine art, paining or sculpture, architecture, music, recorded, or live performance, composition, film and theatre. Finally, it is important as an approach to texts, be they literary in the form of novels, poetry or plays; or be they historical, philosophical, political, legal, moral or religious. Critical theory can be practiced upon any body of written work.

In this section we seek to understand critical theory from its historical roots. We begin by depicting its history and will examine its dawning in the works of German thinkers like Kant, Hegel and Marx, to the present tradition of the Frankfurt School. We will then move on to examine the theoretical components of critical theory. In order to best achieve this, we focus first upon its principal theoretical features as outlined by the *Early* Frankfurt School. We introduce the first formal conceptualisation of the notion of a critical theory with Horkheimer's analysis of critical and traditional theory. Then we move forwards to see an example of a specific critical theory and turn here to that provided by the joint authorship of Horkheimer and Adorno in their seminal *Dialectic of*

Enlightenment. We conclude our section on the Early Frankfurt School by describing Adorno's unusual and remarkable *Negative Dialectics.* Our final task is to approach the formidable work of what has become known as the Later Frankfurt School, the principal figure being of course Jurgen Habermas.

9

The History of Critical Theory

The Frankfurt School's project of critical theory has many historical debts; it would be remiss not to mention a deep past. Directly in tandem with other continental philosophies of the human sciences, critical theory is rooted in traditions, stemming back to concerns shared with the Ancients. Whilst no-one would claim that the Ancients had a type of critical analysis specifically labelled 'critical theory', the Greeks did have an early version of something similar through the figure of Socrates. First, Socratic thought was an attempt to discover truth through reasoned thought. It thereby followed the principles of the later Enlightenment. Truth was best discovered through the power of reason. Secondly, like critical theory, Socratic thought was an attempt to question internally the 'logic' of various arguments and suppositions. It did not look to external definitions for comparison but rather to the 'correctness' of the internal thought procedures themselves. In this vein Socrates developed a 'method' of internal questioning in order to clarify the processes of thought, and Plato depicted the Socratic dialogues and, indeed, first used the idea of 'dialectic'. Whilst the twentieth-century derivations of both internal dialogue and dialectic are somewhat distinct from earlier Greek usage, the concerns of twentieth-century critical thinkers are remarkably similar.

The Ancient Greeks also influenced the Frankfurt School directly. The members of the School for example, were all well-read in the Ancients: in tragedy, in the Greek poets and in the philosophers, Socrates, Plato and Aristotle. Most of Adorno's key texts, whether about aesthetics, epistemology, morality or cultural criticism, are littered with references to the Ancients. For instance, when discussing the relations between morality and wealth, Adorno cites Homer and Aristotle.[1] Moreover, in the English translation of his collection of essays entitled *Critical Models*,[2] there are over eight indexed page references to Plato and seven to Aristotle; similarly in his depiction of aesthetics, the English translation of *Aesthetic Theory*[3] picks out no less than seven page references from Adorno's discussion of Aristotle and twelve references to Plato: compare this with only twelve references to Marx. Yet Adorno's engagement with Marxism is taken as a truism, whilst his humanism is barely recorded.[4] The rich textual discussion of the Ancients continues in his *Moral Philosophy*,[5] wherein Adorno draws heavily upon Socrates, Plato and Aristotle as well as including ample exemplification through Ancient myth. (Even in his *Sociology*,[6] no less, the Ancients get a mention). Finally, Horkheimer and Adorno's seminal critical text, *Dialectic of Enlightenment*,[7] analyses Western society as a fusion, throughout its entire history, of the Ancient and the modern. In order to understand twentieth-century society, Horkheimer and Adorno weigh up the Homeric myths. Odysseus is portrayed throughout the *Dialectic of Enlightenment* as the emblem of twentieth-century man.

Other members of the Frankfurt School besides Horkheimer and Adorno are also fluent in Ancient philosophy. In Marcuse's principal work on epistemological issues, namely, *On the problem of the dialectic*,[8] he begins his work on the notion of reason with a discussion of Plato, not just as a historical reference but in order to actively understand rationality. That is, Marcuse looked to an ancient source as a way of understanding and formulating his own contemporary notion of reason.[9] Indeed, it is hard to find any writings of the principal members of the Frankfurt School that do not richly refer to the Ancients.

Besides engaging with the ancients directly, the entire Frankfurt School's tradition is built upon a philosophy rooted in humanism. Adorno, for example, converses with the Greeks mediated through his

---

[1] 'Since Homer Greek linguistic usage has intertwined the concepts of goodness and wealth . . . Aristotle's Politics openly admits the fusion of inner worth with status in its definition of nobility as inherited wealth, combined with excellence'. Adorno 1974: 184–5.
[2] Adorno 1998.
[3] Adorno 1997.
[4] This is not to claim that Adorno follows or mimics the ancients, his engagement is more often than not critical. It is to argue that he is steeped in a classical education.
[5] Adorno 2000.
[6] Adorno 2000.
[7] Horkheimer and Adorno 1979.
[8] Marcuse 1976.
[9] Held 1980: 229.

German predecessors. For instance, in his epistemological tract *Negative Dialectics*,[10] he quotes Kant's admiration for the Platonic ideal of the Republic ('The Platonic republic has become proverbial as a supposedly striking instance of imagined perfection. . . . One would do better, however, to pursue this thought some more'[11]). More generally, the ideas and issues of the Early Greeks are the bedrock of the major influences upon the Frankfurt School. Kant, Hegel, Nietzsche and Freud were scholars of the Ancients, Kant wrote extensively about Greek philosophy, Hegel too, and we are all familiar with Hegel's work on Greek Tragedy, scattered throughout his *Oeuvre*. Indeed all the German philosophers key to the Frankfurt School both engaged directly with the Greeks and also transmitted humanism indirectly, too.

Critical Theory has roots in the traditions of Judaism and Christianity. Kant was a self-identified Christian and Hegel the son of a Lutheran preacher. Nietzsche sought to engage critically with Christianity, whilst Marx and Freud too were critical of Christian beliefs, but all did so from knowledgeable positions. All, besides being well-versed in the Ancients were equally well-read in the seminal religious transmissions and adaptations of the Classical past.[12] Thus, when we examine influences upon the Frankfurt School, we must remain aware of the Christian tradition-too.

If the Ancient and Christian traditions were conveyed in many general threads and implicit nuances through German philosophy into the modern era, it was with the dawn of Enlightenment that many of the more explicit foundations for Critical Theory were laid. Furthermore, the subsequent developments of the Nineteenth century laid many of the corner stones for the emergence of Critical Theory. Figures like Kant, Hegel, Marx, Weber, Nietzsche and Freud provided the later intellectual mainstay for the Frankfurt School. We can begin our more detailed discussion of specific theoretical points of influence with Kant. Herein, we see elements of Ancient and Christian influences brought to a specially modern context; that of eighteenth-century Enlightenment.

KANT

Immanuel Kant, regarded by some as the world's most distinguished philosopher, lived from 1724 until 1804. He was born in Konigsberg, the fourth of nine children all from parents of a poor harness-maker. His

[10] Adorno 1973.
[11] See Kant, *Kritik der reinen vernunft*, Works III, p. 247 quoted in Adorno, 1973: 330.
[12] Some might say 'adulterations'.

parents were simple people and devout pietists. Kant, from these origins, went on to elaborate a profound philosophy of enlightenment and of duty. He worked and taught in Konigsberg, then in Prussia – now part of Russia. His early works, known as the pre-critical writings were followed by a rest period. There then emerged his three great 'Critiques': the *Critique of Pure Reason*, written in 1781, which dealt with theories about knowledge; the *Critique of Practical Reason*, produced in 1788, which examined ethics and finally, the *Critique of Judgement*, published in 1790, which focussed upon aesthetics.[13]

The German Enlightenment embodied humanist concerns and among other ambitions sought to revitalise the Ancient philosophical preoccupation with the sanctity of reason against the authority of mere superstition or faith. That is, it sought to oppose the strands of Christian tradition with a more rational humanism. Kant, its principal figure, typi-fied the thrust of the epoch with a project that revolved around a critical study of reason and knowledge. It should always be remembered how-ever that Kant was a Christian and that he wanted to marry definitive eighteenth century preoccupations with his commitment to Christianity, that is, to make compatible these competing sets of ideas about the supremacy of either reason or faith.

The seventeenth century had not only itself seen the re-emergence of a concern with reason, it had also seen the emergence of Newtonian Science which had grown and flourished. By the eighteenth century, the Royal Academy had been founded and many of the claims of Newtonian Science had philosophical support in the form of empiricism. Empiricists made claims about the nature of knowledge and asserted that only that which was empirical could count as knowledge. This meant only that which could be experienced through the senses, tested, and verified through experimentation, could be considered knowledge. Science held epistemological supremacy. Against this empiricist view, the rationalists had to vie for authority. Rationalists, against the empiricists, claimed that the world could only be understood through reason. Reason operates not through experimentation but through theorising. Thus, whereas empiri-cists used the methods of science, rationalists turned to the use of the tools of reason, speculative thought and logic. Thereby, they attempted to form theories to represent the world.

Kant was caught between the two schools. He heard the arguments of the empiricists on the one hand, and the rationalists on the other. Each of these schools of thought claimed that they alone held the

[13] See Kant 1996, 1997.

correct position, but Kant was convinced by neither one taken singularly. Taken together, however, he thought that they might *both* be right.

Kant had a second problem. Not only were the empiricists and rationalists in fierce debate, but so, too, were those who used the growing influence of science to undermine the Christian faith. Kant, however, aspired to combine a commitment to enlightenment with Christian faith. He set out therefore, to elaborate a philosophical perspective that allowed empiricism, rationalism and Christianity all to co-exist together without contradiction. This would establish a philosophical unity between competing schools of thought and thereby generate a universal philosophical system.

His ingeniousness in achieving this is not to be understated. Kant divided the nature of the human world into two. He argued that, on the one hand, we have the *real* world which you and I can *observe*. It has trees, rivers, soil, hospitals, schools and human beings in it. This is the real, observable world. To this real observable world we apply all the discoveries of empiricism. We can *know* this world, through empirical observation, scientific tests etc. On the other hand, there exists another 'world'. A world of ideas. These might be ideas about God, or about morality etc., but what all these ideas have in common is that they can not possibly be experienced by the senses. They cannot, therefore, be known either by observation or through experimentation and so we can have no knowledge of them of the empirical kind. However, we can *think* about these things: we can think about God or the nature of morality; we can *reason* about ideas. Rationalism was appropriate to this sphere. Kant, thereby, allowed for the correctness of both empiricism and rationalism by saying that both were correct but each belonged to a different sphere. Empirical knowledge belonged to the real world. Rational theory was the only way to approach the world of ideas.

Finally, Kant accommodated Christianity alongside Enlightenment by placing Christianity in one half of the above schema. Kant placed Christian faith in the world of ideas where everything had to conform to the 'tribunal of reason'. Therefore, by Kant's schema, Christianity had to show rational grounds for its faith, indeed, Christian morality had to be grounded upon reason.

In sum, Kant developed a form of philosophy which brought together empiricism and rationalism with Christianity. He, thereby, generated a system which included various opposing strands of the transmission of humanist ideas. In all, he was the archetypal Enlightenment thinker who grounded his entire philosophical system in reason.

Kant's commitment to the Enlightenment is a strong basis for the Frankfurt School's later developments. Before going on to examine these, it is important to notice, however, that Kant's philosophy was an attempt at a universal philosophy – that is to say, one that is true for all times. Kant, we must remember, was writing against the context of an earlier rapid succession of developments in seventeenth- and eighteenth-century philosophy. Rationalist and empiricist philosophies had competed in turn in these preceding centuries for precedence. His aim, as an eighteenth-century philosopher typifying the thrust of his epoch, was to attempt to construct a universal philosophy which would solve all the earlier dilemmas and *end* once and for all philosophy's *historically variable* character. For Kant, the fundamental nature of the world is invariable over time. Thus, that which is 'real' and observable in the world, and those ideas like the nature of God and morality, do not vary according to their moment in history. Furthermore, the natures of knowledge and reason also do not change in their underlying principles over time. Therefore, although we may gain more and more knowledge, what actually counts as knowledge itself does not change in accordance with our moment in time.

Kant's view of history although in itself rather un-momentous, is significant in that it provides a focal point for a later challenge which was to have huge impact upon German philosophy and subsequently the Frankfurt school. Future challengers in the German tradition overturned the view that knowledge and reason are unchanging over time. The most notable of these contenders was Kant's 'successor' Hegel. Hegel adopted Kant's preoccupation with issues of reason and knowledge and with the Christian faith. He also transmitted ancient ideas into a modern context. He radically dissented, however, from Kant's thoroughgoing a-historicism.

HEGEL

George Wilhelm Friederich Hegel was one of the most important and systematic of the German philosophers to follow after Kant. Hegel developed ideas about knowledge and history that forever changed the scope of European thought. He was born in Stuttgart in 1770. His father was a minor civil servant and other relatives were teachers or Lutheran ministers. He lived a simple life against the backdrop of turbulent and momentous times: the French Revolution. This was also, however, the golden age of German literature and the high point of Romanticism. After flourishing at school, he won a scholarship to Tubingen where he studied philosophy and theology. Subsequently, he went to Jena and joined his friend

Schelling. Later he accepted the post of headmaster at Nuremberg and finally moved on to be Rector of the University of Berlin.

Hegel's point of departure reflects a strong Kantian influence. His motivation like Kant's was to accommodate empiricism with rationalism. However, he wished to elaborate an even more comprehensive Idealist system than Kant had done. Not only did he incorporate the discoveries made by Newtonian Science and rationalism, but also indeed Kant's own supposedly universal and final philosophy. Hegel claimed that, whilst none of these preceding philosophies were wrong as such, they were all *incomplete*.[14] It was he, Hegel, who would develop the first *and final complete* system of philosophy that the world had ever seen.[15]

Hegel's philosophy was, in fact, a direct challenge to Kant's. In contrast with Kant's *Transcendental* Idealism, Hegel developed a system of *Absolute* Idealism with many important distinctions from its predecessor. However, let us not fail to note the important similarities between the goals of these two giants.

Like Kant, Hegel hoped to combine Christian ideas with the emergent importance of Enlightenment rationality. Whereas Kant had made Christian faith compatible with reason by grounding philosophy in reason, Hegel sought to bring Christian faith and reason together by incorporating them both into a *system of history*. That is, he aimed for the same goal as Kant, a project of philosophical unity, but grounded this in a philosophy of history rather than an a-historical system of reason.

Hegel's challenge to Kant, and the feature which set his ideas apart was the completely new philosophical significance that *history* acquired. For Hegel, human history was at the forefront of philosophy and his Absolute Idealism looked for its bedrock to historical idealism.[16]

Christianity

Hegel's system of history was highly indebted to Christianity. Indeed, it was the Christian heritage that actually allowed him to conceptualise a historical journey for the development of reason. We can see that first, from Christianity, the concept of God is central to Hegel. Hegel's idea of God, however, is far from orthodox. His notion of the deity is more humanistic

---

[14] Hegel 1985: 94.

[15] See Hegel 1977, 1985.

[16] There is vast literature on the relation between Kant and Hegel. For some central points see Pippin 1989: 16–41, Rosen, 1982: 115–21 or Walsh in Inwood 1985: 13–30; also Gorland 1996 or Wohlfart 1981.

and indeed 'mystical' than any Christian view. As the distinguished Hegel scholar Walsh noted, 'Hegel's type of philosophy has been denounced as resting on mysticism . . . but there seems no adequate ground for convicting him of [this] defect. He differs from other philosophers not in seeking refuge in mysticism but in taking it seriously'. Walsh continues: 'In his lectures on the history of philosophy Hegel gave Boehme more space than Leibniz, and much more than Hume'.[17] The impact of this humanist mysticism was to develop a conception of the deity as '*Geist*' or 'Spirit'.[18] Geist is a supra-historical entity, embodied in human history, prior to it and consequent of it.

Secondly, Hegel develops a view of history based upon this humanistic, mystical notion of God. Although specifically Lutheran, his notion of the shape of historical development is also developed from general premises derived from the Christian tradition. The basic narrative of creation upon which he based his view is the widespread Christian one that is familiar to us all.[19] This has three parts.

First, in the biblical story of creation, in the beginning, there is an undifferentiated unity of humans with God. This is represented in the idyllic Garden of Paradise. Within the innocent and joyous unity with God, humans have no actual awareness of the possibility of separation from God. This is a perfect unity.

Hegel takes this idea to represent the first stage of an Enlightened human history. He follows this Christian narrative of creation and argues that at the beginning of human, social history, there was indeed a unity. Hegel illustrates this unity with actual historical events – although it should be borne in mind that this is a rather schematic account.

The first stage of human, social history is the unity expressed in myth. Ancient or Classical societies were for Hegel just such mythic societies. Within these, he believed, human beings were simply spontaneously 'at one' or united with the world around them. That is to say, they naturally and unthinkingly behaved in appropriate ways and experienced no disjuncture or conflict with any of the social institutions, rules or forms of authority and knowledge in their society.[20]

[17] Walsh 1985: 29.
[18] For an excellent discussion of 'Geist' see Inwood, 1992: 274–77; Soll, I. 1969 and Solomon, R. 1983.
[19] 'It may be worth mentioning in this connection that Hegel was originally a divinity student, intending to be a pastor in the *Lutheran* Church' (My emphasis), Walsh 1985: 29.
[20] For instance, that depicted in Homer. Hegel 1985: 139–40. See also, for an analysis of approximately the counterpart of the 'logical' stage of development Hegel 1977: 58–103.

Separation

The second stage of development in the Christian story of creation is one of separation. Adam and Eve take fruit from the tree of knowledge and learn shame. This is the Fall. This is how there then occurs a stage of separation. Human beings partake of Original Sin and there is a consequent radical separation of human will from God.

For Hegel, in human history, the second stage, the stage of the Fall, occurs when the mythic unity is disrupted by reflective thought. The philosophy of the Ancient Greek, Socrates, represents for Hegel the beginning of all human thought – when we don't just simply accept the world around us but start *to reflect* upon it.[21] Socratic philosophy leads to a break down of the old communities which depended upon unthinking, habitual, spontaneous forms of behaviour. Hegel writes:

> At the beginning of Greek civilisation, philosophy was tied and bound within the circle of popular religion. Then it extricated itself and took on a hostile attitude to it . . . Plato inveighed against mythology.[22]

There then follows a period of separation, known as *alienation*, in which increasingly thoughtful individuals come into conflict with the social institutions, rules or forms of authority and knowledge in their society. They face social institutions that are increasingly 'other'. Hegel's example of an extreme moment in this process is the advent of the French Revolution, where society critically reflects on the irrational and unjust behaviour of its forms of authority. It then attempts to overthrow these and impose a more rational system.[23]

Reconciliation

In the Christian narrative, after the Fall, there is a long, slow journey back towards reconciliation with God. Reconciliation is a return to unity with God. The possibility of reconciliation is only made possible by the existence of Jesus Christ. He represents the fact that God becomes human in the divine person of Jesus thus demonstrating the possible unity of human and divine.

---

[21] Hegel 1985: 175.

[22] Hegel 1985: 140. See also, for an analysis of approximately the counterpart of the 'logical' stage of development Hegel 1977: 104–262.

[23] See Hardimon 1994: 119–21; Geuss 1995.

History

In Hegel, these Christian ideas become once again incorporated into a view of human history. The final stage of human history, which is actually a long and drawn out stage that encompasses most of human history, is one of reconciliation. In Hegel, reconciliation is the re-unification of humans with the world around them. This happens because the world becomes increasingly rational, and individuals start to recognise this very fact: that their world is becoming increasingly rational. That is to say, their social institutions, rules or forms of authority and knowledge appear to the members of society as rational and just. So the state of reconciliation is when the world is realised by the society within it to be rational. The end point of human history then is the moment of reconciliation which is when, on the one hand, the world attains complete rationality.[24] On the other hand, the rational society recognises itself to be *rational*. Hegel writes: 'The goal, Absolute Knowing' is 'Spirit *that knows itself as Spirit*'.[25]

For Hegel, after the French revolution, the world in Europe becomes more rational. His example of the most rational society was his own Prussian state. Note that this final reconciliation in Hegel is not a return to the original mythic unity where people were unthinkingly 'at one' with their world. In contrast it is a 'return' to a distinct kind of unity wherein people were self-reflective. The world has become rational and yet unity has also returned.

Hegel's theory of the historical development of human society takes the form of initial unity, separation and final reconciliation when the world becomes fully *rational*. The first stage of human history is one without human reason. The second stage is one where humans are rational but in a conflictual way. The third stage is when humans are rational but no longer conflictual as the world around them has also developed to complete rationality and they realise this.

Although Hegel shared the same goal as Kant, the belief that philosophy should be grounded in reason, for Hegel, this is not a goal that can simply be attained by rational thought alone. He believed that for philosophy to become fully rational, human society itself needed to develop to complete rationality. This required the development of human history.

[24] As Hegel expresses it, 'The goal, Absolute Knowing...' Hegel 1977: 493.
[25] Hegel 1977: 493. My emphasis.

In his words: 'Man has needed twenty-three centuries to reach a consciousness of how . . . concept[s] . . . are to be understood'.[26]

The consequence of Hegel's theory of historical and social development is that everything in human life must be understood as part of this historical journey. The entire human world is part of this overall process. For instance, knowledge and understanding develop over time. (Indeed, the main point of Hegel's history is to say that we become more rational through historical development.) Thus, the kinds of knowledge we have and the theories we use change over time. One consequence of this is that we can't use the same understanding from earlier periods as completely true in later times.

Notice how very different from Kant this idea is. Kant wanted to develop a philosophy that was true for all times. Hegel says that Kant's philosophy can't be true for all times because even philosophy is part of a historical process where everything changes.

Hegel was an optimist. He believed that in the end the human world would become rational and just. History has a conclusion, an end point. The historical process achieves its goal. The journey towards a rational unified society of harmony is attained. As Pippin expresses it, Hegel's is a 'detailed history of the education of consciousness . . . to . . . a final completion'.[27]

KARL MARX

The impact of Karl Marx's thought upon critical theory is well-documented.[28] Born in Trier in the German Rhineland in 1818 from a comfortably well-off, moderate, liberal background, his parents were Jews who converted in later life to Protestantism. Marx studied at the University of Bonn transferring to the University of Berlin after a drunken incident and skirmish in a duel. Whilst in Berlin, Marx completed his doctoral dissertation on Ancient themes. During a career in journalism, after marriage and a move to Paris, he began mixing with radicals and socialists. He wrote his enormous volume of works from 1844 onwards, collaborating with Engels, and moving around Europe. From among his many seminal works, those of particular note for critical theorists are his earlier

---

[26] Hegel 1985: 190.
[27] Pippin 1989: 102.
[28] See Bernstein 1992; Geuss 1981; Jameson 1990; Jay 1973, 1984; Kortian 1980.

more philosophical studies including *The German Ideology*,[29] *Economic and Philosophic Manuscripts*,[30] *Communist Manifesto*[31] and *A Contribution to the Critique of Political Economy*.[32]

It was Lenin who declared that the 'three sources and component part's of Marx's philosophy were German philosophy, French Socialism and British political economy.[33] Herein, we are largely concerned with the first: Marx's thought represents a crucial stage of development in the trajectory of German thought from Kant and Hegel to the Frankfurt School.

Marx's motivation was not dissimilar to the Enlightened one of either Kant in the eighteenth, or Hegel in the nineteenth century. Although a secular philosopher whose aim was broadly political rather than religious, he carried forwards a concern with rationality. Moreover, one central component of the expression of his views was his philosophy of history.

Materialism

Famously in his philosophy of history, Marx 'set Hegel on his feet'. He converted Hegel's concern with the ideal realm into a commitment to the more material elements of human existence, specifically, the socio-economic conditions of production. History according to Marx, was not driven by the development of a semi-metaphysical '*Spirit*' but by human labour and the socio-economic relations within which that labour was embedded. He claimed: 'History is the history of human industry, which undergoes growth in productive power, the stimulus and vehicle of which is an economic structure, which perishes when it has stimulated more growth than it can contain'.[34]

History

Although the distinction between Marx and Hegel is clear, it must not be forgotten that Marx adopted many of Hegel's notions about history. First, Marx adopted the same dialectical and teleological model of historical development.

---

[29] Marx 1964.
[30] Marx 1970.
[31] Marx 1992.
[32] Marx 1971.
[33] Lenin in *Marx, Engels, Lenin,* 1972: 452.
[34] See Marx in Cohen 1978: 26.

Unity

Marx's philosophy of history carries forwards the Christian narrative of creation transmitted through Hegel. As with Hegel he theorises a starting point to Western history and generates a semi-speculative model of an original primitive unity. Moreover, he combines this with a Rousseauvian influence: this original primitive unity echoes Rousseau's 'state of nature'.[35] For Marx, in contrast to Hegel, this is a material unity. History begins with the unity of a primitive kind of communism wherein there is no inequality of the form of ownership and non-ownership of the means of production. However, the form of the communism is very undeveloped and the productive powers of this society minimal.

Separation

The second stage of history parallels Hegel's earlier model and entails alienation. The original primitive unity is disrupted and more technologically sophisticated forms of production emerge. These include the feudal model, early industrial capitalism and, later, more and more advanced forms of capitalism. As the means of production grown, socio-economic relations become more dialectically opposed. A rift is generated between the owners of the means of production, on the one hand, and the non-ownership producers, on the other. The latter, infamously, are exploited in Marx's view and the resultant class tension results in a politicisation of the exploited class.[36]

Reconciliation

The final projected stage of history is achieved through political revolution. The politicised proletariat revolt and the institution of private ownership is abolished. The means of production is jointly owned and class inequality becomes a thing of the past. Herein reconciliation is attained in the form of communism. History has developed to a new more sophisticated unity.

Marx offers a similar model of history to Hegel's own. First, he regarded history as having a crucial *generative* role. Communism is brought into existence by the development of the material processes of society. Secondly, history is teleological – a developmental process with an end – moving from primitive unity, through separation to a final reconciliation. Thirdly, historical development occurs dialectically, through the

---

[35] See Kolakowski 1978.
[36] See Marx 1977a.

unravelling of tensions.[37] In general, the tensions in the material forces of society drive historical change; in particular, the clash between the ownership and non-ownership of the means of production in late capitalism. Finally, Marx was also like Hegel, an optimist. History ends with a just society of shared ownership of the means of production.

In spite of the similarities between Marx and Hegel, the distinction, that is the materialist nature of Marx's model, is crucial. This is a major change. First, it entails a radical secularisation: unlike Kant and Hegel before him, Marx is not intent upon retaining Christian faith.

Secondly, Marx removes the foundational role of reason. In its place he instates the material world. Thus, ideas, *reason*, forms of 'consciousness' all *result* from the *material* forces developing over history. This point has major implications for the generation of rationality.

Kant believed in the supremacy of reason as the basis for all human action, justice and faith: reason over mere authority, tradition, habit, belief or superstition. Hegel agreed, but argued that reason was not merely an abstract universal that could be applied at any point in time. It developed historically. Full rationality could only be achieved at the completion of western history. Marx agreed with Hegel but believed that historical development was generated by the material conditions of society. Reason, therefore, is dependent upon these. Rationality is enmeshed in human socio-economic production. This is a major departure from Hegel's view. Only when we have a fully, *materially* developed society (communism) can we also have a fully rational society.

Marx's notion of a historical-material basis to reason became highly influential in later Marxism. Concepts like 'ideology', 'reification' and 'false consciousness' were developed by Marxists to account for incomplete stages of history and their concomitant non-rational thought processes and false beliefs. These later Marxists struggled with analysing the ways in which the non-rational emerged and the forms it took.[38] However, later political events generated a scepticism towards Marx's own thought and provoked a rethinking of the relations between the material conditions of society and the development of forms of reason.

LATE MARXISM

Whereas Marx had predicted a communist revolution, by the later nineteenth century none had emerged. Later Marxists hence sought to

[37] See Marx 1977a and b.
[38] See Jameson 1990.

explain why. To this end there developed several distinct strands of Marxist thought from extreme materialism to more idealist variants. The latter strand proved highly influential in certain ways to the Frankfurt School. The most significant voice of this strand was that of Lukacs.

Lukacs

Georg Lukacs was born in 1885 and lived until 1971. He was Hungarian and joined the Hungarian communist party in 1918. His relationship with communism was a problematic one. He moved between Vienna and Berlin, fleeing the Nazis in 1933, moving to Moscow where he lived until the end of World War II. Eventually, he returned to his native Hungary as a university professor. In part due to his troubled life and career, Lukacs oscillated, for external as well as intellectual reasons, between more materialist and idealist convictions.[39] However, in his seminal *History and Class Consciousness* Lukacs represented a splintering off from the materialist wing of Marxism and developed a new strand of thought.[40]

In *History and Class Consciousness* Lukacs accepted that of central importance to historical change was the growth and development of certain kinds of reason. Thus, historical development was not only caused by tensions in the division of labour – the socio-economic strata – but by alterations in the cultural realm. In shifting his emphasis thus, Lukacs was moving away from a Marxist materialist view of history and reverting to the perspective of his Idealist forebear, Hegel.[41]

It seemed however that, even in his early work, Lukacs's position lay ambivalently somewhere between Marxist Materialist and Kantian/Hegelian convictions. For although pinpointing problems in reason, and indeed marking these as pivotal in the stagnation of history's progress. Lukacs in fact retained a materialist explanation of the problems in the 'superstructure'. He argued that it was conditions in the socio-economic sphere that gave rise to problems in reason. Thus, although reason prevented certain changes in historical development, this form of reason was itself characteristic of certain socio-economic conditions.[42]

More common to a Marxist than a Hegelian view, Lukacs saw the tensions that drove history forwards as based upon socio-economic factors. Tensions surrounded the organisation of the division of labour, that is, they involved class. Also like Marx, Lukacs saw the end point of history

---

[39] See Jacoby 1971; Lichtheim 1961.
[40] Lukacs 1971.
[41] Lukacs 1971: 83.
[42] Lukacs 1971: 41.

as the reconciliation of these class tensions, and as such political.[43] This contrasts with the Hegelian view wherein reconciliation was, in principle, the attainment of rationality of which political structure was a mere part.

Finally, Lukacs followed the Hegelian belief that society's mistaken beliefs resulted specifically from *undeveloped* forms of reason. This 'false consciousness' would be overcome and replaced by 'true consciousness', which consisted in the attainment of complete or full consciousness. Herein, Lukacs follows the Hegelian notion that mistaken knowledge and understanding is the result of partial understanding. Complete truth, perfect rationality, lies within the eventual attainment of the rational whole.

Lukacs Marxist convictions remained entrenched in that he earmarked the proletariat as holding these mistaken beliefs. He thereby argued from the point of view of a specific socio-economic class: it was a particular class that suffered from 'false consciousness'. Full consciousness was important in so far as it enabled communism. It was this that was the eventual goal.

THE EARLY FRANKFURT SCHOOL

All these strands of influence outlined here in this chapter, from the Kantian Enlightenment perspective and the Hegelian-Marxist strand conjoined with other diverse strands of influence from nineteenth-century thinkers such as Max Weber, Nietzsche and Freud. A variety of challenges and contributions to notions of rationality emerged. The Frankfurt School, through its various members, synthesised these whilst retaining a remarkable commitment to Hegelian-Marxism.

The intellectual influences of the Early Frankfurt School were, as we have seen, those of German philosophy Kant, Hegel, Marx and Lukacs. However, we have also mentioned that they included Nietzsche and Freud. Nietzsche had challenged Hegel's enlightened optimism and developed a more pessimistic view of Western history. He perceived, especially in his earlier works, not progress but decline, a loss of our Ancient heritage, cultural degeneration and aesthetic weakness. He looked less to the future for a model of perfection and more to the past – especially to the early non-rational Greek legacy of Tragedy and myth.[44]

[43] Lukacs 1971: 162.
[44] See for example Jay 1973; O'Connor 2004.

Freud meanwhile challenged the notion that reason could be attained through the simple exercise of the faculties as Kant had assumed, or through the development of meta-historical or historical forces, be they ideal or material. He inculcated a shift towards the notion that human psychology played a foundational role in forming reason and shaping society. Moreover, he maintained that the individual human psyche was at core irrational and this irrationality provided a potential obstacle to the attainment of rationality.

These influences conjoined in the Early Frankfurt School's work, perhaps no more so than in the works of Marcuse and Adorno as we shall witness later.[45] However, we can still perceive the strand of Hegelian-Marxist commitment to the centrality of history, and the Kantian project of enlightenment throughout the work of all the Early Frankfurt School members.[46]

Horkheimer

Horkheimer, the director of the Early Frankfurt Institute, was responsible for the specific concept of 'critical theory'. He was born in Stuttgart in 1895, the only son from a very distinguished business family. Both parents were fervent Jews. Although his father was a patron of the arts he inculcated Horkheimer into business at an early age. However, Horkheimer through friendships with Pollock and others was drawn into intellectual life, and restlessly travelled throughout Europe, developing his thoughts and pursuing socialist and philosophical preoccupations. His dissertation written in 1925 was on Kant.[47]

Horkheimer, in fact, followed the Kantian-Hegelian orientated trajectory of development of Marxist ideas that Lukacs had initiated. His arguments were propelled forwards in this same direction because he shared the realisation that had driven Lukacs, namely, that Marx's predictions had not manifested themselves. Historical forces had not developed in the expected direction towards class emancipation.

Horkheimer, in common with other members of the Early Frankfurt School, moved much further away from the materialist concerns of 'orthodox' Marxists than even Lukacs had done. He became almost wholly concerned with cultural/epistemological forces as a means of

---

45 See Whitebook 1995; Sherratt 2002.
46 See Buck-Morss 1977 for an account of the Marxist strand.
47 For more details see Wiggershaus 1994: 41–52.

explaining why history had not achieved reconciliation. Indeed, he almost moved away from class-based modes of analysis altogether. Class after all was a material division of society. Moreover, he lacked conviction that the proletariat was revolutionary. Horkheimer could not see that the progressive movement forwards in history resided in forces concentrated in this particular class.

Horkheimer was influenced by Max Weber, who had seen the cultural conditions accompanying developed capitalism as resulting in bureau-cratisation and the spread of unthinking, repetitive forms of labour. Accompanying this repetitive activity was a correspondingly dull mental condition. To be more precise, Weber believed that developed capital-ism entailed such a huge administrative machinery that it incurred the spread of reductive forms of reasoning in tandem with the accompanying inevitable bureaucratisation process.

Horkheimer conjoined Weber's influence with a strong commitment to forms of late Marxism and looked towards the concept of reason itself as the problematic of social stagnation. He believed that after Enlighten-ment and into modernity, the form of reason that had come to dominate *the whole of society*, not just one particular class, was flawed.[48]

Following Weber, the depleted form of reason of his modern society Horkheimer described as *instrumental reason*. This was a form of reason where people could only think about *means*. For example, I could think about the best way to earn money to buy a bigger, new house, but I could not think about whether buying a new house was the best goal in my life. One of the reasons Horkheimer jettisoned Marxist class based forms of analysis was because he believed the problem of instrumental reason permeated the whole of society, not just the proletariat.[49]

A further change Horkheimer inaugurated was to abandon the Lukacsian-Marxist and indeed Hegelian notion of history as moving towards completion. That is to say, he jettisoned the idea that history was a total process moving towards some kind of final goal, for instance, communism. He was not optimistic. He no longer regarded history as necessarily follow a trajectory of development towards a positive conclusion.[50]

Horkheimer in line with the tenet of the early Frankfurt Institute still examined society from a historical perspective. He didn't abandon the

[48] See Horkheimer 1968b.
[49] See Horkheimer 1974.
[50] See Horkheimer 1968b: 145.

idea that society changes over time. Moreover, he accepted the idea of an aim within history. It is simply that, writing against the context of Nazism's spread in Germany, he no longer believed historical development would lead to inevitable emancipation. He was pessimistic. However, it is crucial to note that he did not simply accept that society was not developing at all, that our institutions and rationality were merely stagnating. In pinpointing the problem of society's stagnation, in focussing upon instrumental reason as the central problematic, he also generated the kernel of a solution. Horkheimer developed the idea for a way to overcome instrumental reason. A way to rekindle a sense of purpose in human society, rather than simply resigning himself to the possibility that human life had become one vast, efficient, meaningless and dangerous machine. It was to this end that he generated the idea of a critical theory.[51]

Through critical theory, Horkheimer believed, we could start to think beyond the merely instrumental and reassess the issue of purpose. If not in a strictly positive form, we could at least begin to think critically about what we should aim for, about social values, and most importantly about how and why political barbarism was emerging in Europe. Critical Theory was to be a new kind of reasoning. It was to be a reflective, ends orientated and emancipatory form of reason.[52]

[51] See Horkheimer 1974.
[52] Horkheimer 1974.

10

Critical Theory of the Early Frankfurt School

Horkheimer built his notion of critical theory upon an idea of theory that differed from the ordinary use of the term. He argued that 'theory' was not about developing a string of connected statements that represented the nature of the world around us. He did not believe that the word 'theory' should simply refer to a static set of propositions. In contrast, he thought that theory referred to a kind of *activity* generated within a particular social context. For him *theory* means in essence what you and I would normally take *theorising* to mean.[1]

Human societies are constantly acting upon nature in order to maintain their existence and in order to reproduce themselves. We burn fuel, build houses, plough land, all to perpetuate our human lives. In the course of doing these things, we need to understand the nature of our world so we gain knowledge in order to enable us to produce and reproduce more efficiently. The kinds of knowledge that we develop whilst undergoing these activities of human social and economic survival are processed together as kinds of theories. For example, theories about diseases, about animal migration, geological strata or indeed Newtonian science. All these theories that are part of the activity of human survival are called 'traditional' theories, according to Horkheimer.

Horkheimer believed that there was a very different kind of theorising to the traditional variant. This different type of theorising had quite distinct goals from the former. Whereas traditional theories were aimed at efficient survival, a new kind of theory could be developed that was aimed at transforming society to become more rational. Indeed, a new

[1] This is a summary of a much more complex argument taken from Horkheimer 1968a.

kind of theorising could be developed aimed at helping society progress in the direction of Enlightenment. This would also be emancipatory.[2]

Horkheimer called this new kind of theorising *critical theory*. Critical theory was aimed at both enlightenment and emancipation. Critical theory was not, however, 'positive' in the sense that it offered statements about the nature of Enlightenment, rationality or emancipation. It was not a set of propositions that told us how to live our lives. This would simply be an instance of the old kind of theory: static and historically insensitive. Critical theory was oppositional, indeed, as its name suggests, 'critical'. It could not tell us what an enlightened society looked at, rather it showed us how our own society failed to live up to being enlightened. Critical theory pinpointed our delusions, irrationality and those of our society. Likewise, critical theory could not provide a statement of linked propositions about what emancipation in our society would look like. It could show us, however, how our own society failed to be emancipated. Critical theory could reveal to us that our world failed to achieve its own ideals including that of emancipation.[3]

Critical theory proceeds through a very particular and distinctive kind of process known as *immanent* or *internal critique*. This approach is inspired by Hegel's historicism. Let us pause for a moment to see how.

Hegel, as we have seen, believed that human activities including social institutions, traditions etc., are directed towards an inherent goal. Rather as an acorn is a potential oak tree, so social institutions and all forms of human behaviour are potential ideal forms of themselves, according to Hegel. For example, a moral child might be a potentially perfect priest; the foundations of a building are potentially a new industry; a lazy and corrupt institution is potentially an efficient and effective hospital. In fact, if we recall Hegel's optimism about society, we are all progressing towards eventual enlightenment when we live in a just and rational society. Just as society is progressing towards enlightenment as a whole, so every aspect of human life is progressing towards its own perfection. Like an acorn, within every aspect of society is the kernel of this perfection which will later emerge.

A further idea exists within Hegel's social philosophy. He believed that the kernel of perfection, or aim, within every aspect of human society could, in a certain way, be known. Adorno captured Hegel's idea of knowing the inner perfection in things when he pointed out that it was

---

[2] Horkheimer 1968a.
[3] Horkheimer 1968a.

as if Hegel were claiming an institution could 'speak' about itself. If an institution were to describe itself it would define itself in accord with its own aim. For example, the aim of a hospital would be that it cured all sick people. In practice, the hospital staff might be lazy or inefficient. Consultants might be more concerned with financial rewards than with curing the sick. Bureaucrats might be more obsessed with petty procedures than being genuinely efficient. Administrators might be more concerned with following government procedures and regulations than with really getting the hospital functioning properly. In all the day-to-day procedures the individuals lose a sense of the overall aim of the institution. At this point the institution can be described as a 'hospital', for it has officials, doctors, a budget, buildings, equipment, etc. But it is far from being an ideal hospital. However, the kernel of an aim is somehow embedded within the institution itself. So, if the hospital could speak, and you were to ask the hospital what its aim was, it would say, 'to cure the sick': within the very concept 'hospital' lies this ideal, its aim.

Institutions thus have two aspects to them. On the one hand they have lots of observable aspects which we can describe and collate facts about. Hospitals have corridors, personnel, operating theatres etc. On the other hand, they have within them an *internal aim*. Hospitals aim to cure the sick. Furthermore, this internal aim is embedded within the very definition of the institution itself.

The nature of this *internal aim* is far from subjective. It is not what you or I think or imagine a hospital is, it is what the hospital *itself* aims to be. So we have three ways of depicting a hospital. First, there is an external, factual description of the building, the staff, the way the hospital runs etc. Second, there is your or my personal opinion about the hospital, whether it is good or bad, ugly or just our view about what it does. Third, there is the internal aim of the hospital, what the institution's goal is. This internal aim is embedded within the very definition of the hospital itself.

Critical theory is interested primarily in the third point, the internal aim of the institution. This is the definition of what the institution believes itself to be, which includes its own aim. How critical theory works is to try to recover that aim, show the difference between what an institution's original purpose is and what it actually is at the present time. That is to say, critical theory presents an institution with its own pretensions about itself and thereby reveals a discrepancy between what it aims to be and actually is (at that point in time).

The process of confronting an item with the reality about itself in contrast with that which it aims to be is known as '*internal criticism*'. This internal criticism is the essence of critical theory. For example, it might reveal the discrepancy between the aim of curing the sick and the fact of consultants demanding ever-increasing amounts of pay. It might confront the hospital with the discrepancy between curing the sick and the reality of long waiting lists for patients. It shows the vast amounts of money going into a complex and inefficient administration. Note that this criticism is not *my* criticising the hospital for failing to be what you, I or a third party think it should be. (A government might think hospitals should be converted to businesses as it would save public funds and might mean they could reduce taxes and thereby gain more votes.[4])

Critical theory, therefore, works by confronting social institutions, practices or indeed individuals with the discrepancy between their own *internal* aim and the actual state of their affairs. In this way, it criticises an item according to that item's own internal *standard* and *definition* of itself. This contrasts with the notion of external criticism. In external criticism, an individual or body of individuals criticise something according to what *they* think that institution should be. For example, a Catholic might criticise a liberal for using contraceptives. It would simply be an instance of an external criticism because liberals don't have any aim which includes not using contraceptives. This would be an external Catholic criticism of a liberal. However, if a Catholic used contraception, we could offer an internal criticism of their conduct because part of being a Catholic is a set of beliefs and aims which include not using contraception. Another example might be, if a judge were to accept a bribe. This could be criticised internally because the office of a judge includes the aim of being 'just', which includes not accepting bribes. Therefore, to accept a bribe is to fail to live up to the internal aim of being a judge. Finally, consider the example of a father criticising a son who is a composer for not earning much money. This would be an external criticism, because the son's aim in being a composer revolves around composing high-quality music, not making profit *per se*. The aim of amassing wealth is external to that of being a composer. In short, internal criticism must capture the internal aim of an institution, office or form of practice.

This process of revealing the discrepancy between an internal aim and the actual reality of an item is the process of critical theory. Critical theory, in revealing these discrepancies, aims to push society forwards in the direction of emancipation and enlightenment. It aims to help us think critically about the society we live in and uncover all forms of irrationality,

4 Horkheimer 1968a.

whether these be delusions, hypocrisies or ideologies. Whether they are the consciously manipulated propaganda of obnoxious political regimes, the hypocrisy of corrupt institutions or the hypnotic degeneracy of the mass media, Hollywood or advertising.[5] By generating this critical theorising, an emancipatory practice is unleashed to engender awareness as to the true nature of the social items around us: a vision of the discrepancy between the purpose of the social world and what it has become, thereby leaving space for individuals to reunite institutions and practices with these institution's own, internal aims.

We can witness two examples of critical theory – note that these are not Horkheimer's own instances but they are influential theories which contain the essential features of critical theory. Highly influential in the nineteenth and twentieth centuries are Marxism and psychoanalysis.

In Marxism, Marx develops a view of history wherein Western history has an internal aim. It aims for equality and rationality of the material conditions of life. When this socio-economic equality and rationality is obtained, the other features of human society can then also become rational and just. Thus, according to Marx, history has its own, internal aim; the aim of 'enlightenment' understood from a Marxist perspective.

Material 'enlightenment' occurs in the form of communism. This is Western history's own internal aim. Note, Marx does not view communism as his own subjective preference but as that of Western history. Communism in his view is history's own internal standard with which we can compare all actual historical occurrences. In so far as social, political and economic events aim towards this greater aim of history, they are rational. In so far as they do not, they are irrational. Social, political and economic action, can therefore, according to Marx, be judged according to the internal aims of history itself. We can, therefore, practice internal criticism of social events and practices. We can practice critical theory.

Moreover, Marxism is itself a critical theory. Marx presents a view of history wherein phenomena like feudalism and capitalism can be judged according to the overall aim of history, the realisation of communism. We can analyse capitalism in so far as it lives up to the aim of history – achieving communism. We can criticise its discrepancy from this aim – failing to achieve communism.

Psychoanalysis works in a similar way.[6] In psychoanalysis, individuals are considered to have certain aims and objectives which are impeded

---

[5] See Horkheimer 1972.
[6] See Habermas 1971: Chapters 10–12, who also uses this as an example.

by their various neuroses and forms of abnormality. However, individuals cling to all these neurotic traits and generate all sorts of delusions about themselves and others. Psychoanalysis seeks to reveal to individuals the discrepancy between what they aim to be and how they are actually being. In so doing, psychoanalysis aims to emancipate and enlighten individuals about themselves.

Psychoanalysis is also a form of critical theory within itself. Psychoanalysis has the aim of emancipation and enlightenment. It aims to help people become normal, healthy individuals who can live fulfilling lives. In so far as psychoanalysis fails to live up to its own aims, it can itself be criticised. Thus psychoanalysis both works through a process of critical theory and can also itself be subjected to critical theory.

Dialectic of Enlightenment

Having viewed nineteenth-century philosophies which have the traits of critical theory, it remains for us to depict one of the Early Frankfurt School's own Critical Theories. The thesis we are going to inspect is arguably the most prestigious Critical Theory of the Early Frankfurt School. This is the extremely ambitious study of the entire history of Western society: a Critical Theory of Western history. Its aim was to enlighten society about itself. Indeed, as the Early Frankfurt School regarded their own society as an instance of 'enlightenment', their study was aimed, in their own words, at 'enlightening the enlightenment about itself'.[7]

The critical theory of which we speak, namely, *Dialectic of Enlightenment*, was jointly authored by Horkheimer and Adorno. They each took responsibility for every word. After witnessing the rise of fascism in Europe and fleeing the terrors of Nazism during World War II, Horkheimer and Adorno wrote *Dialectic of Enlightenment* to address the most pertinent question of their times: why had twentieth-century Western society, a product in their view of enlightenment, degenerated to Nazism? Why had the society of their native Germany, with the aims of Enlightenment, Reason, Knowledge, Freedom, Peace, Stability, Progress resulted in a society which embraced Ignorance, Domination, Brutality, War and Regression?

Enlightenment

In order to see how Horkheimer and Adorno answered this question, we need first to clarify their use of the concept 'enlightenment'. Historians

---

[7] Horkheimer and Adorno 1979: Introduction.

and humanities scholars in general use the term Enlightenment to refer more or less to the eighteenth century and the culture of reason (often regarded in opposition to faith). Rationality as the basis of politics, law, knowledge, action, ethics etc. is deemed the predominant characteristic of Enlightenment. Horkheimer and Adorno certainly accept this view. However, they see the characteristic of Enlightenment as spreading well beyond the eighteenth century and into their own contemporary society. The reason for this is that they view history as embodying an aim. Thus for them, enlightenment refers to a society with a series of aims which they believed permeated not only the eighteenth century but continued into their own day.

For Horkheimer and Adorno, therefore, enlightenment is the same as historical Enlightenment in that it refers to the same set of ideas as historical Enlightenment. They write: 'Enlightenment, according to Kant is . . . man's emergence from his self-incurred immaturity. Immaturity is the inability to use one's understanding without the guidance of another person'. 'Understanding without the guidance of another person is understanding guided by reason'.[8] Their concept of enlightenment is distinct in that it is philosophical not historical. Moreover, enlightenment for them is a speculative not an empirical notion. 'Enlightenment' referred to a cultural trend they believed underlay their own Western society.

Enlightenment, for Horkheimer and Adorno, was defined according to its aims, the principal one of which was the acquisition of knowledge. The acquisition of knowledge was linked to the Subject's attainment of maturity and to a set of further aims namely, freedom, security and peace – all of which constitute, for the enlightenment, progress.[9]

The defining of enlightenment according to its aims is the first feature of critical theory. By defining enlightenment according to its aims, Horkheimer and Adorno are able to practice *internal critique*. (Remember that internal critique means that one notices the discrepancy between what an item *aims* to be – which is also what it *claims* to be – and what it has in fact *become*). They explain: 'The point is rather that the Enlightenment must examine itself'.[10]

Horkheimer and Adorno's project is an analysis of how and why the enlightenment fails to live up to its aims. It is important at the outset

[8] Horkheimer and Adorno 1979: 81.
[9] Horkheimer and Adorno 1979: 3, 81.
[10] Horkheimer and Adorno 1979: xv.

to understand that they find *no fault* with any of the enlightenment's own *aims*. Indeed the point of their criticism is precisely that the enlightenment *fails* to achieve these, which is to say, they fully endorse these aspirations:

> In the most general sense of progressive thought, the Enlightenment has always aimed at liberating men from fear and establishing their sovereignty. Yet the fully enlightened earth radiates disaster triumphant. The programme of the world; the dissolution of myths and the substitution of knowledge for fancy. Bacon, the father of experimental philosophy, had defined its motives. He looked down on the masters of tradition, the great reputed authors' who first 'believe that others know that which they know not; and after themselves know that which they know not'.[11]

Horkheimer and Adorno's criticism is internal. They judge the enlightenment according to its *own* standards and do *not* introduce any *external* standards whatsoever. They simply assess the enlightenment in so far as it manages or does not manage to realise what it sets out to achieve. Their project is to rescue the enlightenment from its own failure. Their goal of criticising enlightenment is aimed at emancipation. The emancipation of enlightenment from fascism coupled to the emancipation of their own contemporary society – to free this from Nazism.

Let us look now at the details of their critical theory. Enlightenment aims for Reason, Knowledge, Freedom, Peace, Stability and Progress. In mid-twentieth-century Germany it had declined to Nazism with the features of Ignorance, Domination, Brutality, War and Regression. How? Horkheimer and Adorno argue that the problems of twentieth century Germany have been present throughout Western history. Nazism was not an isolated phenomenon. It was not an anomaly. Nazism was intrinsic to the project of enlightenment itself.

How can that be? How can Nazism be bound with enlightenment which in every sense of its aims is the complete opposite. Enlightenment, Horkheimer and Adorno believe, is indeed the opposite of Nazism. Enlightenment aims to be the opposite of Nazism, but fails. One of the reasons for this is that Enlightenment and Nazism are related. They are related dialectically.

The term 'dialectical' refers to a specific kind of relationship between two items. This relationship has two particular characteristics. First, it refers to two items being inextricably linked. This means they cannot be separated. A second characteristic of a dialectical relationship is that

[11] Horkheimer and Adorno 1979: 3.

two items are oppositional. The two items are related to each other in an antagonistic fashion. Many people think that because two items are oppositional, they cannot be joined. But this is precisely the point.[12]

Nazism and enlightenment cannot be separated. Moreover, they are oppositional. The point that Nazism and enlightenment are oppositional may come as no surprise to us, but the view that they are inextricably linked must surely be rather surprising. To explain this latter point, we need to see that for Horkheimer and Adorno, Nazism is a specific instance of a much larger phenomenon.

Myth

These Early Frankfurt School members believe that Nazism is an instance of myth. From their perspective enlightenment, culture has a view about itself which is that enlightenment is the complete opposite of (what it regards as) another kind of '*culture*', namely myth. Horkheimer and Adorno write:

the program of the enlightenment was the disenchantment of the world; the dissolution of myths and the substitution of knowledge for fancy.[13]

Their concept of myth like enlightenment, is rather unusual. As a starting point to understanding their notion of myth, it should be confused neither with any notion of a non-Western, 'primitive' culture, nor with a literary genre. Their notion of myth is derived from their interpretation of 'classical' or 'ancient' Western culture. However, classical Western culture, as it is generally understood, is a broader and more complex culture than is captured in their concept of myth. They would regard classical Western culture as containing non-mythic elements, for instance, elements of enlightenment itself. Furthermore, they would consider it to possess, in general, traits that are both inherently positive and elements that are inherently negative. Their own concept of myth is, however, wholly negative. It is an ideal type based upon certain negative traits that they perceive in classical Western culture. They often elucidated this concept through an interpretation of Homer.

Myth, for Horkheimer and Adorno is permeated by a certain set of attitudes which differ from the enlightenment in that they are *not*

---

[12] Imagine that a married man and woman are inseparable but squabble all the time. This would be a good metaphor for a dialectical relationship.

[13] Horkheimer and Adorno 1979: 3.

derived from a set of aims. In their view, myth does not have any aims at all and simply 'is what it is'. It is, basically speaking, centred around a way of relating to the world which is 'animistic'. This involves a particular system of knowledge acquisition for which they deploy the term *animism.*

Animism is a kind of *projection.* Horkheimer and Adorno explain, 'The separation of the animate and the inanimate, the occupation of certain places by demons and deities, first arises from this pre-animism, which contains the first lines of the separation of subject and object. When the tree is no longer approached merely as tree, but as evidence for an Other, as the location of *mana,* language expresses the contradiction that something is itself and at one and the same time something other than itself... Through the deity, language is transformed from tautology to language'.[14]

What Horkheimer and Adorno are explaining is that animistic projection occurs when, instead of seeing the world 'as it is' we project our fears and desires onto it. We imagine demons and monsters, or fairytale princes. The early Frankfurt School members' twentieth century parallel examples are the Nazi projections about the nature of the Jews, the demonising of that which is different, the inculcation of fear of the Other. Nazi ideas about the Super race of Arians, moreover, bear many resemblances to fantastical projections about 'princes'.

Horkheimer and Adorno regard mythic projection as a 'false' system of knowledge acquisition. This false knowledge can relate to the further traits of myth which are immaturity, social domination, an expression of fear and barbarism.

Enlightenment Fails

The enlightenment sees itself as having transcended myth; as having overcome myth's *negative* features of animism, immaturity, domination, fear, barbarism and regression. According to Horkheimer and Adorno, the entire self-conception of enlightenment is, in fact, formed in opposition to myth. They believe that enlightenment fails and that this failure is of the nature of a regression to myth. The regression of enlightenment into myth is what the enlightenment itself would conceive of as a regression into its absolute opposite and thus a sign of complete failure. It is a regression that encompasses *all* aspects of enlightenment culture.

---

[14] Horkheimer and Adorno 1979: 15.

Dialectic of Enlightenment is a speculative theory of Western history. Historically speaking, myth and enlightenment have been present in some
sense ever since the dawn of Western culture. Adorno and Horkheimer
write: 'myth is already enlightenment; and enlightenment reverts to
mythology'.[15] Moreover, they were both present in mid-twentieth-century
Germany. The whole of western history is a narrative of the dialectic
between enlightenment and myth. Enlightenment is the opposite of myth
and is constantly trying to conquer myth and realise its own aims of reason, peace, security, freedom etc. However, due to this inextricable link
with its opposite, enlightenment can never sever itself from myth and
therefore always remains likely to collapse into myth, into barbarism and
domination.

Western history in fact is a series of oscillations between enlightenment
and myth. Sometimes enlightenment is the predominant force, but at
other times, myth gains the upper hand. Both are always present. Nazism
can be explained as a time in which myth came to predominate over
enlightenment.

Horkheimer and Adorno offer further details of enlightenment and
myth. They argue that both these cultures are driven to cognitive and
reflective activity by a drive for survival, and an accompanying fear of the
external world. In myth, this drive takes on a more primitive shape. In
myth, peoples imitate the 'threatening other' in order to appease it.[16]
They dress up as threatening animals, they symbolise the elements in
dress and ritual, they represent natural forces and disasters with various
Gods. They then try to make themselves like these gods, in order to be
secure from the god's wrath. Furthermore, they offer sacrifices to ensure
their safety.

In enlightenment, Horkheimer and Adorno argue, we have supposedly transcended these 'primitive' forms of behaviour. However, we have
our own 'myths'. The myths of Nazism and the astrological *'hocus pocus'*
of the United States are some of Adorno and Horkheimer's prime examples. But their central example of myth was the bedrock of Western culture itself: a scientific and bureaucratic culture. Our fear of nature and
the 'Other' displays itself in our obsession for control. Through science
we attempt to make nature like us, like our systems and theories and
technological practices. In converting nature to be 'like us', we hope to
overcome its threatening 'Otherness'. Whereas primitive mythic peoples

[15] Horkheimer and Adorno 1979: xvi.
[16] Horkheimer and Adorno 1979: 15, 31.

imitated nature, we make nature imitate us. The drive is the same fear for survival.

Moreover, just as primitive peoples made sacrifices, so, too, do we. We invent vast bureaucratic systems which we take to be 'rational'. But these are really just expressions of fear. Through having these great bureaucracies, modern peoples attempt to track and control the minute details of life. This activity is driven by fear, just as primitive, mythic sacrifice was. Moreover, within these bureaucratic systems exist repressive and controlling forms of behaviour which represent a sacrifice of human spontaneity and pleasure. We do not douse the corn with a blood sacrifice to ensure that it grows, but we douse our institutions in the sacrifice of our human blood, understood as our human 'spirit', in order to ensure our own survival.

We have seen how Horkheimer and Adorno produce a philosophy of western history that is a critical theory of enlightenment. Western history aims to attain the same aims as enlightenment. These aims include being the opposite of myth, completely transcending myth. Therefore, to show how enlightenment still includes many features of myth is to enlighten the enlightenment about itself. That is, *Dialectic of Enlightenment* produces an internal critique of enlightenment showing how enlightenment fails to live up to its own internal aims.

11

Innovations in Critical Theory

Adorno and Habermas

If Horkheimer viewed critical theory as twentieth-century Western society's source of redemption, Adorno looked for another solution. Theodor Weisengrund Adorno was born in Frankfurt am Main on 11 September 1903 to a family with exotic ancestral background, including a Corsican fencing master for a grandfather, from whom the name 'Adorno' derived. (His real family name being Weisengrund which, with its Jewish overtones, was substituted for 'Adorno' during his exile in the United States in order to avoid possible anti-Semitic prejudice). He was a precocious and intellectually formidable young man communicating with a variety of the most prestigious musical and philosophical minds of Twentieth Century Germany including Kracauer, Berg, Schoenberg, Benjamin, Brecht, Bloch, Horkheimer and later Mann, among others. Besides inspiring his musical talents, these intellectual friends and mentors spread the ideas of Kant, Hegel, Nietzsche, Marx and other canonical philosophers. During the rise of the Nazis in Germany, Adorno followed his Jewish colleagues into exile in 1934: he was but thirty-one years of age. In the United States, Adorno was exposed to a new culture, sociological research and his relationship with Horkheimer deepened. They co-authored *Dialectic of Enlightenment*, and Adorno penned many other key texts. He returned to Germany in 1949 and was finally recognised as an eminent public figure. His interests encompassed psychoanalysis, television research, analysis of horoscopes and prophecy, literary criticism, the making of radio broadcasts, the writing of many of his key philosophical texts including the Marxist-Idealist inspired epistemological study *Negative Dialectics*, and the aphoristic *Minima Moralia* which triumphed to become a best seller. Adorno died during the student uprising of the 1960s during which he was tragically heckled by his lecture audiences.

NEGATIVE DIALECTICS

Many of Adorno's works have contributed to philosophy of social science. His jointly authored text with Horkheimer, *Dialectic of Enlightenment*, is perhaps central. However, if this represents a historical expression of critique, Adorno's singly authored *Negative Dialectics* is an epistemological variant of similar standing.

Throughout his career Adorno believed that the activity of acquiring knowledge was inseparable from other activities like social and, indeed, emotional ones. However, in spite of being adamantly against the notion of epistemology as a separate discipline, Adorno generated his fairly focussed analysis of knowledge acquisition. In this, he attempted to go beyond the idea of a critical theory and generate a kind of knowledge acquisition that was itself inherently critical. That is to say, he went beyond Horkheimer's conception of a critical theory as a distinct kind of theoretical practice. Moreover, he went beyond the idea of generating a specific critical theory, like the example of *Dialectic of Enlightenment*. Adorno generated a form of acquiring knowledge that embedded critique within itself. That is to say, he sought to establish a way of knowing that within its very own constitution was self-critical. This was his idea of *Negative Dialectics*.[1]

There are, broadly, speaking, three main ways of conceptualising Adorno's idea of negative dialectics. All revolve around the notion of *non-identity thinking*. First, therefore, let us assess this concept.

Adorno believed that *identity* thinking was the process involved in all conventional kinds of cognition and theorising. Knowledge proceeds through a process of categorisation and identification. The subject perceives an object and then identifies it through a concept, for instance, a small, dark, furry object might be identified by the concept 'cat'. In this identifying form of organizing activity, objects become assimilated into our man-made categories. In pursuing this procedure, we lose something of the object itself. That is to say, our categories fail to capture some meaning of the object in itself. Although identity thinking is deemed necessary – indeed unavoidable – if we wish to know the world around us, it is inherently limited.[2]

---

[1] Adorno 1973.
[2] Adorno 1973: 134–210.

The problem of the restricted nature of identity thinking is exacerbated in that we tend to believe we have identified objects completely with our concepts. One result of this is that we lose a sense of the possibility of a concepts' incompleteness or possible inaccuracy. For exemplification's purpose we can compare an identifying category with a photograph of an object. A photograph is always a particular, indeed partial representation of an object. However, if we take the photograph to represent the object in its entirety, we lose a sense of the potential for other un-photographed elements in the object.

A further consequence of identity thinking is that our concepts come to replace the actual object. That is, we stop looking at the Object and become more concerned with its representation. That is to say, in identity thinking we not only take these 'photographs' of objects to be the object but we also actually supplant the 'real' object with these very photos. We then mistake the photo for the real object and this in turn is a form of delusion. Adorno refers to this process as reification.[3] This reification of our own identificatory tools, our classification system, leads to an inherently limited and delusionary kind of knowledge acquisition. It is to remedy these ills of identity thinking that the notion of non-identity thinking is developed.[4]

Non-identity thinking has several aims. On the one hand, it aims to make the subject aware of the limits of his or her own knowledge. In fact (less subjectively), it aims to make the concept itself aware of the limits of the concept. Adorno tries to develop a form of knowledge acquisition where the concept negates itself in the very process of identification. An example of this kind of non-identity thinking can be found within his own work. His discussions always entail statements and counter statements, claims and their oppositions: a language of paradox. This way of theorising is an attempt, neither to confuse the reader, as it sometimes seems, nor to promote a kind of pseudo-mythical thinking but to set concepts against their oppositional meaning so that identities are always negated. To offer an example, Adorno discusses works of art and writes about the concept of 'the heterogeneous' within art. He says 'What is heterogeneous to artworks is immanent to them'. So far this seems fairly straightforward. But then he writes: 'it is that in them that opposes unity' and he continues in the same sentence to claim that heterogeneity is also that, in turn, which 'is needed by unity'. The whole sentence thus

[3] A Lukacsian-derived concept.
[4] Adorno 1973: 134–210.

reads: 'What is heterogeneous to artworks is immanent to them: it is that in them that opposes unity yet it is needed by unity'.[5] In this statement the concept 'the heterogeneous' has two contradictory meanings. On the one hand it is that element within art that opposes the unity of the overall work of art. On the other hand, the heterogeneous is that element which is needed by unity. Adorno's notion of identifying 'unity' as a feature of an artwork is contradicted by negating 'unity' as a feature of an artwork. Adorno's aim in non-identity thinking is to proceed to define items, like art, with conceptual features and then to immediately negate these self-same definitions in order to ward off the idea of any inherent reification. Adorno believes that through this constant negation of identification that he can develop a form of thought that does not become complacent and assume it has captured the object in its totality. Furthermore, that thought is prevented from becoming a static representation but forced, through these oppositions to be ever dynamic and ever embracing new contradictions.

Non-identity thinking has several features. First, it is an attempt at immanent criticism of the form of thought itself. Second, it is also an attempt to go beyond the concept and to reach out to the object itself. This second feature of non-identity thinking takes us into a realm of semi-mystical thinking within Adorno's work. It is a realm which many describe as indebted to Walter Benjamin. (Although my own view is that there is a certain similarity between Benjamin and Adorno here, I would be inclined to attribute this to an intuitive affinity between these two thinkers rather than an (external) influence of one upon the other).

Adorno's semi-mystical notions about non-identity thinking, stem from deep within his own artistic sensibility.[6] To go beyond the concept and reach the Object itself, entails, according to Adorno, to go beyond language itself. It is the outer limit of the concept, the furthest point of extension of the process of identification that reveals to us the outer edge of our concepts (and hence bears much similarity to the Kantian sublime). The point where language is strained to its limit and disappears into silence. At this point, the human voice vanishes and there is a temporary glimmer of the 'voice' of the object. It is a poetic moment. The moment, for example, when the music disperses and we have only silence. Herein the object can gleam in its own inherent identity for a fractional moment. For this reason, Adorno's own style of theorising is

[5] Adorno 1997: 89.
[6] Adorno 1973: 134–210.

both contradictory and poetical. It is an attempt to define and contradict the hard identities of classificatory thought. It is also an attempt to go beyond them and let the object speak for itself in the poetic glimmer of silence between the noises of identification.[7]

Non-identity thinking of this second silent variant, is also a negation. It is a negation of the image of the concept by a non-image, a moment of absence. It is also the negation of the noise of the concept by a moment of silence.

In short, negation occurs through an oppositional identity to a positive identity and by an absence of an identity to an identity present. Negative dialectics proceeds through contradiction and absence. Both these are attempts to undermine the reificatory nature of consistent, rational, positive, identificatory thought.[8]

If we recall that Adorno wrote against the background of the rise of Nazism in Germany and fled to exile in the United States, we can begin to understand the influences behind these rather obscure ideas. First, there is a deep history in German thought stemming from the Enlightenment and beyond, wherein rationality is believed to be positive, in every sense including moral – Kant for instance. A strong current of anti-Semitic thinking in German philosophy centres around the idea of Jews as irrational. Rationalism has been deemed to be something inherently missing in Judaism.[9]

Further, Nazi ideas were often equated with the dominance of science and technology and their associated forms of rationality. Adorno, as an intellectual with Jewish ancestry, was encountering a prejudice about Jews being irrational and a conception of reason that was linked to techno-scientific domination. He was, therefore, strongly motivated to find a form of rationalism in what had been rejected by the dominant Enlightenment and scientific-technological culture as irrational. It was not that Jews were irrational, it was that the 'external' standards of rationality they were measured by, were themselves irrational.

In point of fact, Adorno went further and argued that in certain respects Enlightenment itself was irrational! Enlightenment, after all, advocated positive identity thinking in all its realms, cognitive and theoretical, real and ideal. And all these led to reification and concomitant irrationality.[10]

---

[7] Adorno 1973: 134–210; see also Adorno 1997.
[8] Adorno 1973: 3–60.
[9] See Mack 2003.
[10] See Horkheimer and Adorno 1979.

Adorno sought to oppose the dominance of positive identity thinking. *Negative Dialectics* is both an analysis of identity and non-identity thinking and also an instance of the latter. After all, if Kant advocated systematic identity thinking as the basis of reason we would expect him to write his text in what he believed to be a rational way. Adorno has done the same, except that his form of reasoning is a challenge to the positive identificatory one that is deeply embedded within the Western tradition.

(Note, however, that Adorno is no post-modernist. He is not an advocate of any kind of nominalism. In contrast to Nietzsche – in his later work – Adorno believes objects are 'out there' with their own inherent identities. He is in fact a strange kind of 'realist'. It is our ability to know those identities which are suspect, not the objects themselves. We could perhaps best encapsulate Adorno as a 'mystical realist' on the one hand, and a rational idealist, on the other.[11])

In spite of earlier optimism, however, in the further years of the development of the Frankfurt School, critical theory, in all its forms became antiquated. The members of the early Frankfurt School were traumatised by the events of Nazism and then disappointed with the culture of the United States to which they had fled. In both Nazi Germany and in the United States, the Frankfurt School thought that they were presented with societies where cultures lacked any fundamental purpose. These societies, institutions and practices simply had no aims. In fact, efficient functioning was their only 'attribute', which, of course, is the same 'aim' as survival – that is, a mere means, not a proper (moral, meaningful) purpose itself. These country's institutions did not aim for any Good, for any morality or enlightenment, the members of the Early Frankfurt School, believed. They were content to be merely what they were, efficient, functional machines of wealth. The result of this, the critical theorists resigned themselves to accepting, was that there was no possibility of internal criticism. Without aims, there was no room for a discrepancy between the aim and the actuality. Hitler, in spite of supposed mythic ideals, in actual fact had no perverse (moral) aims, merely instrumental domination. Thus when he condemned the Jews, how could he be *internally* criticised. Likewise, America sought only the proliferation of efficient wealth. How could the culture industry be *internally* criticised for failing to enlighten or uphold aesthetic standards when it merely served to efficiently produce and distribute a commodity? Critical theory became obsolete.[12]

[11] See Sherratt 2002.
[12] See Held 1980.

Notice that in all this the Early Frankfurt School followed the view that 'hypocrisy' was a higher stage of enlightenment than simply being without any aims at all. If something failed to live up to its aspirations, it could at least be criticised. Moreover, there was a certain sophistication in having aims. However, when something had no (normative) aims, it was content merely to be an efficient functioning machine, an adept beast, there was no place for criticism at all. Such an item was at a very primitive stage of development. Indeed, societies without aims had almost reverted back into nature – there was barely any standard to denote them as civilisations at all.

HABERMAS

Jürgen Habermas was born in Germany in 1929. His childhood was spent against the backdrop of Nazi Germany. He reached maturity well after its demise and became an academic teacher first in Heidelberg, then Frankfurt, where he held a professorial Chair before moving to the Max Planck Institute in Starnberg. He came under the influence of the Early Frankfurt School and was taught by Theodor Adorno, later becoming his assistant. His main works include *Structural Transformation of the Public Sphere*,[13] *Towards a Rational Society*;[14] *Knowledge and Human Interests*,[15] *Legitimation Crisis*,[16] *Theory and Practice*;[17] *Communication* and *the Evolution of Society*.[18]

His intellectual influences hold much in common with the Early Frankfurt School, spanning from Ancient Greek philosophy through to the canonical thinkers of the German tradition as outlined above. In common with the ancients, his project involved a focus upon 'true thinking' and like them too, he was interested in a form of knowledge that combined 'fact' and 'value'. He was deeply troubled by scientific, 'technocratic' forms of thought spreading at an alarming rate through his contemporary society. Although inheriting ideas from Romanticism and from Hegel, Nietzsche and Freud, he was perhaps the member of the Frankfurt School to be most strongly influenced by 'rational' Enlightenment thinkers and materialists too, two of his main inspirations being Kant and

[13] See the translation in Habermas 1989.
[14] Habermas 1970.
[15] Habermas 1971.
[16] Habermas 1976.
[17] Habermas 1974.
[18] Habermas 1979.

Marx. Habermas borrowed from a plethora of other, later social thinkers such as Karl-Otto Apel, Mead, Goffman, Parsons, Piaget and Kohlberg.[19]

His oeuvre is formidable and far too complex to summarise here. It stems however from arguably the central preoccupation of the Early Frankfurt School, namely the role of reason in society. One of the principal issues uniting all his work is that of how knowledge is related to human social activity.

In his early work, *Knowledge and Human Interests*, Habermas explores the socio-economic basis to capitalism and concludes, in true Weberian fashion, that this entails widespread bureaucratisation. A shallow, deficient kind of reasoning accompanies bureaucratisation, a means/ends rationality wherein individuals do not ponder in any meaningful way the multiple facets of their lives. A repetitive mode of thinking whereby they simply ascertain the most efficient means to pursue any given ends is all that this social mode of production entails. This kind of 'instrumental reason' spreads in modern capitalist societies, almost a disease of the mind, for Habermas. This bureaucratic culture, as much as any form of neurological disorder, generates a deficient mentality in its populace. Individuals can no longer think about the purpose of human actions and lives.

Furthermore, in *Knowledge and Human Interests*[20] and in his later work *Theory and Practice*,[21] Habermas continues to look to Marx more than to Kant. It is the historically generated socio-economic conditions that he believes determine 'the structure of objects of possible experience'. However, Habermas includes a far greater emphasis upon language than Marx had ever done. Knowledge, for Habermas is built upon the twin pillars of material labour and human language.

He develops complex categorisation systems to denote the relative roles of labour and language in reason. Human beings, he argues, organise knowledge based upon their interests. These are firstly, productive, that is, *technical interests*. Secondly, communicative, that is, in Kantian derived language, *practical interests*. Thirdly we have self-reflective, self-determining, that is *emancipatory interests*.[22]

These interests unfold in a variety of social activates. Work is built upon productive interests and entails instrumental kinds of action.

[19] For more details, see White 1995.
[20] Habermas 1971.
[21] Habermas 1974.
[22] Habermas 1971.

Practical interests are involved in social interaction and this revolves around the use of both language and power. Power generates asymmetrical relations of dependency and constraint with implications for the third set of interests – namely, emancipatory.[23]

Three forms of rational activity emerge from these different interests and social activities. First, mainly useful for work are the empirical-analytic modes of thought. Secondly, concerned with language and human interaction are the historical-hermeneutic skills. Thirdly, concerned with self reflection and human emancipation are critical modes of reasoning.[24]

Habermas complicates this tripartite structure somewhat by illuminating a thread that occurs throughout all domains of reasoning. Most knowledge and reasoning occurs, he believes, in situations of distortion. Human interests primarily in the technical and practical domain entail power relations. These result in the distortion of the truth. Thereby, knowledge and reason become inflected by ideology.[25]

Habermas's aim is for genuine rationality free from the distorting effects of ideology. This goal is what he sets out to achieve in his project of a critical theory. His argument runs as follows. All speech, he claims, aims towards consensus. He refers to the conditions in which this aim is freely pursued – free, that is, from coercion and its effects – as the 'ideal speech situation'. Under such circumstances as the ideal speech situation the kind of consensus attained is a 'rational consensus'. However, to attain a rational consensus requires more than simply avoiding the obvious effects of coercion. It entails avoiding the distorted communication that power relations evoke, namely, all corresponding forms of ideology. The transcendence of these systems of distorted communication, that is, ideology, occurs through self-reflection. This is a critical kind of reflection and is the building block for his notion of critical theory.[26]

Habermas's conceptions are, as you can see, a far cry from Adorno's semi-mystical negative dialectics. Habermas has returned in fact to a more Kantian style of criticism. He has borrowed Kant's notion of the 'transcendental' realm, namely, that beyond observable experience. Herein, Habermas has argued, we can find normative aims embedded within speech. It is in this semi-Kantian conception of the 'ideal speech' situation that Habermas invests his optimism.

[23] Habermas 1971.
[24] Habermas 1971, 1974.
[25] Habermas 1971, 1974.
[26] Habermas 1971, 1974.

Because Habermas's notion of critical theory is indebted to an underlying system of social analysis, much of his work is also focussed upon these social analyses. Besides critical theory his work included a critique of ideology, social systems analysis and evolutionary theory (incorporating ontogenesis and phylogenesis). His later works also involve detailed social, economic and political as well as more strictly 'critical' analyses of capitalism. A major component of Habermas critical theory is an analysis of instrumental reason that involves a pretty thorough reworking of Marx.

In contrast to Marx, Habermas focuses upon the importance of communication over material labour in human relations. In one of his best-known works, on the *Structural Transformation of the Public Sphere*,[27] he pursues this conviction about the importance of linguistic interaction by generating the notion of the necessity of a 'free' communicative realm in human society. His work centres around the concept of the 'public sphere' which he defines as 'a realm of social life in which something approaching public opinion can be formed'. In common with his other early ideas, his notion of a realm in which a genuine public opinion can be formed, is of one which is free from coercion, authority, dogma or any other distorting power relations. It is only in such a power-free environment that the standards of critical reason proper to the development of public opinion can be upheld.[28]

His analysis is in the main historical. He argues that the public sphere developed during a phase of early capitalism. Its role was to mediate between state and society. Its importance was that it was the very foundation for any civil society. However, he laments that the increasing complexity and scale of later capitalism, the need for management, the administration of the sheer mass of the populace and its resources, delivers a shrinkage of the public sphere. This realm of public opinion is all but lost and bureaucracy becomes the new preoccupation of government, and the new (unthinking) mediating force between state and society.

These preoccupations are clearly present throughout his *oeuvre*. In both *Towards a Rational Society*[29] and *Theory and Practice*,[30] Habermas laments the loss of the public sphere and the deterioration of reason accompanying later capitalism. In later capitalism, commerce has

[27] Habermas 1989.
[28] Habermas 1989.
[29] Habermas 1970.
[30] Habermas 1974.

encroached, he argues, upon the activities of the public sphere. Bureaucracy generates instrumental reason. Science replaces purposive, meaningful kinds of discourse and reason. The state is forced into an inter-
ventionist role in order to protect capitalism, all of which results in the
reduction and compression of the public sphere.

The results Habermas argues of these social changes accompanying
capitalism are catastrophic for rationality. First, there occurs a shift away
from distorting ideologies to an actual narrowing of the form of reason.
It is not that individuals hold false beliefs but that they are denied the
very skills of reasoning to hold any beliefs at all! The critical reason of
the public sphere vanishes.

Secondly, this deterioration in the capacity to reason Habermas argues,
is paradoxically a form of ideology in itself and it serves the interests of
later capitalism.

Thirdly, the role of science and technology with their accompanying narrow forms of reason spread. This *technocratic reason* becomes used
to solve social, political and moral issues as well as the economic and
technological problems for which they were designed. (This situation
as described by Habermas can be compared to the rationality of asking
a computer programme how we should live a meaningful life.) Technocratic reason spreads everywhere and in fact becomes a more pervasive ideology than any previous one. Its expression can be found in
approaches to the humanities like positivism, rational choice theory, political economy as a mode of social analysis and all analytic forms of reason.
Habermas refers to all these through the powerful conception of '*technocratic ideology*'.

In the language of the categories described in Habermas's earlier
works, we could sum up these points by saying that *work rationality* in
later capitalism spreads into the sphere of human *communication* and
emancipation, severely depleting these latter.

Habermas's solution is a form of critical theory which encompasses
both his own critical analyses of societies, which are in themselves demon-
strations of critical reasoning; it also entails the vastly ambitious project of
re-working Marxism. Habermas declared that he wanted to rework Marxism to be immune to assimilation into technocratic ideology. In order
to do this, he stresses that language and inter-subjectivity should replace
the concept of the 'macro-subject' of historical materialism. This macro-
subject itself bares some of the traits of technocratic ideology. In order
to achieve this, Habermas borrows from the philosophy of language.
From speech-act theory he adopts the distinction between strategic and

communicative thinking. The key trait of communicative thinking being that it is criticisable. Reasoning then has a role. A theory cannot simply be true or false in the way that an empirical fact can be.

In short, Habermas argues that critical theory is associated with communication and emancipation. It is a form of knowledge that serves the human interest of freedom: it is related to 'the conduct of one's life'. In common with earlier versions, he pursues the notion of emancipatory, cognitive interest and emphasises autonomy. His distinct contribution is the focus upon speech and consensus.

Although Habermas is perhaps the most influential critical theorist in contemporary philosophy, his actual concept of critical theory treads little novel ground. His own analysis often bares the categorisational mode of thought typical of instrumental reason and his lack of analysis of human emotions and the importance of the aesthetic realm leave him popular within the very analytic tradition which he apparently abhorred. Adorno's *Negative Dialectics* is a far more profound alternative to technocratic ideology, undermining its dearest assumptions. The most valuable part of Habermas's analysis is perhaps his critique of the fabric of late capitalism and his historical work on the public sphere which is undoubtably masterful.

Conclusion

Our first aim in *Continental Philosophy of Social Science* has been to provide a much-needed specialist study of European traditions of analysis of society, which have often been marginalized in favour of the more mainstream Anglo-American approaches. We have focussed upon arguably the principal strands within the continental rubric, namely hermeneutics, genealogy and critical theory. Demonstrating the distinctness of these traditions from the Anglo-American science-dominated agenda has been our ambition throughout. To this end we have concentrated on discussing the continental tradition's own canon of thinkers, questions, style of analysis and, moreover, its humanist origins.

The second purpose of our study has been to argue for and explore in detail the specifically humanist nature of the continental tradition. To this end we have defined humanism and showed how each tradition embodies the principal features of humanism. Our definition entailed three essential features. First, we claimed that humanism builds upon a knowledge of the Ancients: that it is an approach to learning based upon thorough scholarly engagement with the Greek and Roman authors of antiquity. Second, we argue that humanism entails a conception of knowledge as transmitted through the ages. Thirdly, we claimed that humanists believed the world to be meaningful and further that that meaning is created to the greatest extent by human beings. We then demonstrated the humanist character of each of our three principal traditions of hermeneutics, genealogy and critical theory.

Of all the traditions in continental philosophy of social science, hermeneutics is perhaps the foremost instance of humanist endeavour. We have tried to show that it is not possible to fully appreciate the hermeneutic tradition, past and present, without having a full sense of its deep entrenchment within humanism. We have argued that hermeneutics vividly displays all three features of humanism. It is rooted in the Ancients. Indeed, hermeneutics was first practiced in Greece and Rome and owes much of its character to these early beginnings. Secondly, hermeneutics works through the transmission of knowledge. It maintains a living connection with its Ancient past. At every stage in its historical development from the Early Church Fathers through to the Enlightenment and into the present, hermeneuticists have been well educated in Ancient texts and traditions. Some of the most influential recent figures, Heidegger and Gadamer, for example, produced scholarship steeped in references to and understanding of the ancient inheritance.

Thirdly, hermeneutics throughout all its twists and turns construes the human world as meaningful. The whole tradition is dedicated to interpreting and understanding that meaning, which is conceptualised as either created by human historical activity and/or through human linguistic action. Throughout the course of its development hermeneutics has pursued the idea of the importance to humans of both their historical past and their faculty for speech, albeit with differing emphases in the work of the Romantics, German existentialists and more contemporary figures.

In common with hermeneutics, we have seen how genealogy also displays the essential humanist features. It is based upon Nietzsche who was, of course, an eminent classicist and intimidatingly well read in the Ancients. His ideas were then taken up and the ancient knowledge transmitted to Foucault. In his deployment of Nietzsche's philosophical perspective, Foucault carries many humanist assumptions and inspirations into a highly contemporary context.

Although Nietzsche and Foucault are often regarded as radical and as challenging many traditional assumptions, we have seen how, in fact, their work carries forwards the main foundation of humanism, namely, the notion that the world is meaningful and that human beings create this meaning. We have seen how in his early works Nietzsche construes meaning as residing in artistic experience and as generated through human aesthetic engagement. In his later work on genealogy this emphasis upon the humanly constructed nature of meaning continues. History through contingency and power struggles constructs meaning. This meaning changes over time and so words, concepts, ideas and expressions are not historically invariable. Indeed, the very identity of the objects around us changes during the course of time.

To claim that meaning is created through power and contingency, whilst a radical version of humanism, and one that is quite at odds with certain earlier humanist and especially Christian-humanist preconceptions, is not at all at odds with the overall tenets of humanism. Many have interpreted Nietzsche's idea that power and contingency are responsible for the creation of meaning as suggesting, almost, that the world is in fact meaningless. That is to say, that because meanings are constructed in arbitrary and non-intentional ways, that Nietzsche is arguing for the non-existence of any kind of 'real' meaning. This would be a deeply misleading view. Power struggles, whilst perhaps not the most attractive element of human activity are certainly derived from *human activity*. Furthermore, whilst not intentional, *contingency* in Nietzsche's genealogy is certainly the contingency arising out of *human activity*. Thus, both the rather unflattering vision of power struggles and the rather opaque notion of contingency are both kinds of human action. Thus, it is human action, in Nietzsche's view, however mediated, that generates history and creates meanings. The radical historical nominalism of Nietzsche, therefore, is clearly in fact humanist.

Likewise, the social constructivism of Foucault, whilst challenging many foundationalist principles, also carries forwards the flame of humanism. Indeed, Foucault grounds science itself in humanist maxims. In his view, the human sciences regime is historically generated through human power struggles and activity. Whilst Foucault views the truth and meaning that humans create as not necessarily always inherently good or indeed sincere, he does, however, devotedly follow the humanist maxim that meaning is created by human beings. Our world is meaningful, and indeed, our very truths are created through historical activity.

Critical theory, too, we have demonstrated, is a thoroughgoing humanist tradition. Almost the entire Frankfurt School read the Ancients. Indeed, their key text *Dialectic of Enlightenment* is a critical theory built upon a reading of Homer's Odyssey. The commonly acknowledged eighteenth-century foundations to critical theory are also those of philosophers well read in the Ancients. Moreover, the Christian forebears to the German Enlightenment thinkers we have shown to be transmitters of the Ancient heritage.

Critical theory is in a very real sense a transmission of knowledge from the ancient past. Indeed, critical theory is best understood as a reworking of the ancient notion of substantive reason for contemporary times. Critical theory seeks to apply a negative version of the much admired substantive reason of the Ancients to an era of the twentieth century.

Critical theory is also dedicated to a humanist notion of meaning. For the members of the Frankfurt School, meaning is never prior to human beings or somehow external to them. It is created by human historical activity – be that material or ideal, the fusion of labour and cultural activity. Reason is perhaps the most essential aspect for them of the human historical creation of meaning. Moreover, the Frankfurt School's project is not merely one of transmitting humanist ideas and ideals but of reinstating them. They argue, too, that not only do human beings create meaning but they also destroy it. Indeed, the social world is more or less meaningful depending upon the sophistication of human activity at any point in time. The Frankfurt School's warning, during their own era, was that although we had inherited a meaningful world, with the growth of instrumental activity and accompanying irrational myth, we were in jeopardy of losing that.

In short, we have attempted to show in *Continental Philosophy of Social Science* that the continental rubric is an important, rich tradition which addresses many contemporary issues, and transmits many important dimensions of knowledge from the past. Most importantly, we have tried to emphasise the humanist approach to the study of society and highlight that this is a vibrant tradition, one which challenges many assumptions of the scientific-led approaches and also that guards against many reductionist assumptions prevalent in the social sciences. We have emphasised throughout the strong, autonomous and historically deep tradition of *Continental Philosophy of Social Science* against a more superficial Anglo-American view that this is a purely ancillary rebellion against science or analytic analysis. Finally, we have attempted to preserve an awareness of the historical depth of the Continental tradition for those who work and deploy its ideas in modern contexts, for it is, ironically, the very disciplines that most use continental philosophy of social science that are often most in danger of doing it an injustice. Sociologists, political scientists, human geographers, literary critics, lawyers, even historiographers and continental philosophers might study Gadamer, Foucault or Derrida, for instance, pay scant attention to deeper traditions and, therefore, sometimes misconstrue these thinkers and their works.

Bibliography

Adorno, T. 1973. *Negative Dialectics*. Trans. E. B. Ashton. London: Routledge and Kegan Paul.

Adorno, T. 1974. *Minima Moralia: Reflections from Damaged Life*. Trans. E. Jephcott. London: New York: Verso.

Adorno, T. 1997. *Aesthetic Theory*. Trans. R. Hullot-Kentor. Minneapolis: University of Minnesota Press.

Adorno, T. 1998. *Critical Models: Interventions and Catchwords*. New York: Columbia University Press.

Adorno, T. 2000. *Problems of Moral Philosophy*. Cambridge: Polity Press.

Adorno, T. 2000. *Introduction to Sociology*. Cambridge: Polity Press.

Agassi, J.; Jarvie, I. C.; Laor, N. 1995. *Critical Rationalism, the Social Sciences and the Humanities*. Dordrecht: Kluwer Academic Publishers.

Apel, Karl Otto. 1971. et al. eds. *Hermeneutik und Ideoloiekritik*. Frankfurt: Suhrkamp.

Anderson, R. J.; Hughes, J. A.; Sharrock, W. W. 1986. *Philosophy and the Human Sciences*. Croom Helm: London and Sydney.

Ansell-Pearson, K. 1996. *Nietzsche contra Rousseau: A Study of Nietzsche's Moral and Political Thought*. Cambridge: Cambridge University Press.

Aquinus, St. Thomas. 1981. *Summa Theologica*. Westminster: Christian Classics.

Aristotle 1853. *The Organon*. London: H. G. Bohn.

Aschheim, S. E. 1992. *The Nietzsche Legacy in Germany, 1890–1990*. Berkeley: University of California Press.

Augustine, St. Bishop of Hippo. 1995. *De Doctrina Christiana*. Oxford: Clarendon Press.

Azevedo, J. 1997. *Mapping Reality: an Evolutionary Realist Methodology for the Natural and Social Sciences*. New York: SUNY Press.

Babette, B. E. and Cohen, R. S. 1999. *Nietzsche, Theories of Knowledge and Critical Theory*. Dordrecht, London: Kluwer Academic Press.

Babich, B. E.; Cohen, R. S. 1999. *Nietzsche, Epistemology and Philosophy of Science* Dordrecht, London: Kluwer Academic Press.

Babich, B. E.; Bergoffen, D. B.; Glynn, S. 1995. *Continental and Postmodern Perspectives in the Philosophy of Science*. Avebury Press.

Barnes, B. 1974. *Scientific Knowledge and Sociological Thinking*. London: Routledge and Kegan Paul.

Benn, S. I.; Mortimore, G. 1976. *Rationality and the Social Sciences: Contributions to the Philosophy and Methodology of the Social Sciences*. Routledge.

Benton, T.; Craib, I. 2001. *Philosophy of Social Science: The Philosophical Foundations of Social Thought*. Hampshire: Palgrave.

Berlin, I. 1976. *Vico and Herder* London: Hogarth.

Berlin, I. 1989. *Against the Current: Essays in the History of Ideas*. Oxford: Clarendon.

Bernstein, J. 1992. *The Fate of Art: Aesthetic Alienation from Kant to Derrida and Adorno*. Cambridge: Polity Press.

Betti, Emilio 1962. *Die Hermeneutik als Allegmeine Methode der Geisteswissenschaften* Tübigen: J.C.B. Mohr.

Bhaskar, R. 1978. *A Realist Theory of Science*. (2nd edn.) Brighton: Harvester.

Binswanger, L. 1963. *Being-in-the-World. Selected Papers of Ludwig Binswanger*. Trans. Needleman, J. New York, London: Basic Books.

Bleicher, J. 1980. *Contemporary Hermeneutics. Hermeneutics as Method, Philosophy and Critique*. London: Routledge and Kegan Paul.

Bleicher, J. 1982. *The Hermeneutic Imagination: Outline of a Positive Critique of Scientism and Sociology*. London: Boston and Henley.

Bloor, D. 1974. 'Popper's Mystification of Objective Knowledge.' *Science Studies*, IV, pp. 65–76.

Bloor, D. 1976. *Knowledge and Social Imagery*. London: Routledge and Kegan Paul.

Bohman, J. 1991. *New Philosophy of Social Science: Problems of Indeterminacy*. Polity Press.

Bottomore, T. 1974. *Marxist Sociology*. London: Macmillan.

Bruns, Gerald L. 1992. *Hermeneutics, Ancient and Modern*. New Haven: Yale University Press.

Bryant, C. 1985. *Positivism in Social Theory and Research*. London: Macmillan.

Buck-Morss, S. 1977. *The Origin of Negative Dialectics: Theodor Adorno, Walter Benjamin and the Frankfurt Institute*. Hassocks, Sussex: Harvester Press.

Bultmann, R. 1955. *Essays Philosophical and Theological*. Trans. Greig J. C. G. London: SCM Press; New York: Macmillan.

Bultmann, R. 1957. *History and Eschatology: The Gifford Lectures, 1955*. University Press, Edinburgh.

Bultmann, R. 1958. *Jesus and the Word*. New York, London: Scribner's.

Bultmann, R. 1958. *Jesus Christ and Mythology*. New York, London: Scribner's.

Bultmann, R. 1983. *Theology of the New Testament*. 2 Vols. London: SCM Press.

Bultmann, R. 1985. *New Testament and Mythology and Other Basic Writings*. Ed. Trans., Ogden, S. London: SCM Press.

Calvin, Jean. 1845. *Institutes of the Christian Religion: A New Translation*. Edinburgh: Calvin Translation Society.

Chladenius, J. M. 1985 (1710–1759). *Allgemeine Geschichtswissenschaft*. Einleitung, Friederich, C. Wien. Böhlau.

Chladenius, J. M. 1969. *Einleitung zur richtigen Auslegung vernünftiger Reden und Schriften.* Intr. Geldsetzer, L. Vol. 5. Series hermeneutica, Instrumenta Philosophica. Düsseldorf: Stern Verlag.

Cicero, M. T. 1942. *Brutus and De Oratore.* Cambridge, MA: Harvard University Press; Heinemann Ltd.

Cohen, G. A. 1972. 'Karl Marx' and the Withering Away of Social Science' *Philosophy and Public Affairs,* 1,2 (Winter).

Cohen, G. A. 1978. *Karl Marx's Theory of History: A Defence.* Oxford: Clarendon Press.

Cohen, I. B. 1993. *The Natural and Social Sciences: Some Critical and Historical Perspectives.* Kluwer Academic Publishers: Dordrecht, Boston.

Cohen, R. S.; Wartofsky, M. W. eds. 1983. *Epistemology, Methodology, and the Social Sciences.* Dordrecht: D. Reidel Publishing Company.

Copeland, Rita. 1991. *Rhetoric, Hermeneutics and Translation in the Middle Ages* Cambridge: Cambridge University Press.

Craib, I. 1992. *Modern Social Theory: From Parsons to Habermas.* Sussex: Harvester, Wheatsheaf.

Craib, I. 1997. *Classical Social Theory.* Oxford: Oxford University Press.

Critchley, S. 2001. *Continental Philosophy: A Very Short Introduction.* Oxford: Oxford University Press.

Dannhauser, K. 1630. *The Idea of a Good Interpretation*

Derrida, J. 1976. Trans. Spivak, G. C. *Of Grammatology.* London and Baltimore.

Dilthey, Wilhelm. 1914–1990. *Gesammelte Schriften,* 20 vols. Vols. 1–12, Suttgart: B. G. Teubner, and Gottingen: Vandenhoeck & Ruprecht; vols. 13–20, Gottingen: Vandenhoeck & Ruprecht.

Dilthey, Wilhelm. 1985–. *Selected Works,* 6 vols. Trans. Makkreel, R. A. and Rodi, F. Princeton, NJ: Princeton University Press.

Dilthey, Wilhelm. 1989. *Introduction to the Human Sciences.* Ed., Intro. Makkreel, R. A. and Rodi, F. Princeton, NJ: Princeton University Press.

Dilthey, Wilhelm. 2002. *Formation of the Historical World* in *the Human Sciences.* Ed., Intro. Makkreel, R. A. and Rodi, F. Princeton, NJ: Princeton University Press.

Doyal, L. 1986. *Empiricism, Explanation and Rationality: An Introduction to the Philosophy of the Social Sciences.* London: Routledge & Kegan Paul.

Doyal, L.; Harris, R. 1986. *Empiricism, Explanation and Rationality: an Introduction to the Philosophy of the Social Sciences.* London: Routledge.

Dray, W. H. 1957. *Laws and Explanation in History.* London: Oxford University Press.

Dray, W. H. 1959. "Explaining what is History." In *Theories of History.* Gardiner, P. ed. Glencoe, IL: The Free Press.

Droysen, Johann Gustav. 1877. *Geschichte Alexanders des Grossen.* Berlin: R. V. Becker.

Droysen, Johann Gustav. 1897. *Outline of the Principles of History.* Trans. Andrews, E. B. Boston: Ginn and Co.

Droysen, Johann Gustav. 1917. *Geschichte des Hellenismus.* Gotha: Friederich Andreas Berthes.

Droysen, Johann Gustav 1977. *Historik: Vorlesungen uber Enzyklopadie und Methodologie der Geschichte.* Ed. Hubner, R. 8th. ed. Munich: R. Oldenbourg.

Dupre, J. 2003. *Darwin's Legacy. What Evolution Means Today.* Oxford: Oxford University Press.

Ebeling, G. 1959. *"Hermeneutik." Religion in Geschichte und Gegenwart.* 3rd ed.: 242–46.

Eden, Kathy. 1997. *Hermeneutics and the Rhetorical Tradition: Chapters in the Ancient Legacy and its Humanist Reception.* New Haven, London: Yale University Press.

Ernesti, J. H. 1699. *Compendium Hermeneutiae Profanae.* Leipzig: np.

Fay, B. 1996. *Contemporary Philosophy of Social Science: A Multicultural Approach.* Oxford: Blackwell.

Ferguson, D. S. 1986. *Biblical Hermeneutics,* London: SCM.

Feyerband, 1962. 'Explanation, Reduction and Empiricism' in Feigl, H. and Maxwell, G. eds. *Minnesota Studies in the Philosophy of Science, Vol. 3.* Minneapolis: University of Minnesota Press.

Flew, A. 1985. *Thinking about Social Thinking: The Philosophy of the Social Sciences.* Oxford: Blackwell.

Forstman, J. 1977. *A Romantic Triangle: Schleiermacher and Early German Romanticism.* Missoula, MT: Scholars Press.

Foucault, M. 1977. *Language, Counter-memory, Practice: Selected Essays and Interviews by Michel Foucault.* Bouchard, ed. Ithaca, NY: Cornell University Press.

Foucault, M. 1978. *The History of Sexuality: An Introduction.* Harmondsworth: Penguin Books.

Foucault, M. 1979. *Discipline and Punish: The Birth of The Prison.* Harmondsworth: Penguin Books.

Foucault, M. 1980. *Power/Knowledge.* Hassocks, Sussex: Harvester Press.

Foucault, M. 1985. *The Use of Pleasure: The History of Sexuality Volume 2.* Harmondsworth: Penguin Books.

Fuchs, E. 1959–65. *Gesammelte Aufsätze.* 3bd. Tubingen. 1959–65.

Gadamer, H. G. 1981. *Reason in the Age of Science.* Cambridge, MA and London: MIT Press.

Gadamer, H. G. 1989. *Truth and Method.* Trans. Weinsheimer, J. and Marshall D. G. London: Sheed and Ward.

Galvotti, M. C. 2003. *Observation and experiment in the natural and social sciences.* Dordrecht: Kluwer.

Gavroglu, K.; Stachel, J.; Wartofsky, M. W. 1995. *Science, Politics and Social Practice: Essays on Marxism and Science, Philosophy of Culture and the Social Sciences* Dordrecht: Kluwer Acadmeic publishers.

Geertz, C. 1973. *The Interpretation of Cultures.* London: Fontana.

Geertz, C. 1988. *Works and Lives: The Anthropologist as Author.* Cambridge: Cambridge University Press.

Geertz, C. 1993. *Local Knowledge. Further Essays in Interpretive Anthropology.* London: Fontana.

Geuss, R. 1981. *The Idea of a Critical Theory: Habermas and the Frankfurt School.* Cambridge: Cambridge University Press.

Geuss, R. 1995. Cambridge Lecture Series on Hegel's Philosophy of History, unpublished.

Geuss, R. 1999. 'Nietzsche and Genealogy', in his *Morality, Culture and History.* Cambridge: Cambridge University Press.

Glassner, B.; Moreno, J. D. 1989. *The Qualitative-Quantitative Distinction in the Social Sciences.* Dordrecht: Kluwer.

Glynn, S. 1986. *European Philosophy and the Human and Social Sciences.* Aldershot: Gower.

Goldman, A. I. 1992. *Liaisons: Philosophy Meets the Cognitive and Social Sciences.* Cambridge, MA: MIT Press.

Golomb, J. 1997. *Nietzsche and Jewish Culture.* London: Routledge.

Gordon, L. R. 1995. *Fanon and the Crisis of European Man: An Essay on Philosophy and the Human Sciences.* London, New York: Routledge.

Gordon, S. 1991. *The History and Philosophy of Social Science.* London, New York: Routledge.

Görland, I. 1996. *Die Kant Kritik des Jungen Hegel.* Frankfurt: Klostermann.

Habermas, J. 1970. *Towards a Rational Society.* Trans. Shapiro, J. J. London: Heineman.

Habermas, J. 1971. *Knowledge and Human Interests.* Trans. Shapiro, J. J. London: Heinemann. (Habermas, J. 1986 *Erkenntnis und Interesse* Frankfurt)

Habermas, J. 1974. *Theory and Practice.* London: Heinemann.

Habermas, J. 1976. *Legitimation Crisis.* Trans. McCarthy T. London: Heineman. (Habermas, J. 1973 *Legitimationsprobleme im Spätkapitalismus* Frankfurt)

Habermas, J. 1979. *Communication and the Evolution of Society.* Trans. McCarthy T. London: Heineman. (Habermas, J. 1981 *Theorie des kommunikativen Handelns*, 2 vols. Frankfurt)

Habermas, J. 1982. *Zur Logik der Sozialwissenschaften.* Frankfurt: Suhrkamp.

Habermas, J. 1989. *The Structural Transformation of the Public Sphere.* Cambridge: Polity Press.

Halfpenny, P. 1982. *Positivism and Sociology.* London: Allen and Unwin.

Hardimon, M. O. 1994. *Hegel's Social Philosophy: The Project of Reconciliation.* Cambridge: Cambridge University Press.

Harrison, R. ed. 1979. *Rational Action: Studies in Philosophy and Social Science.* Cambridge: Cambridge University Press.

Hayek F. A. 1948. *Individualism and Economic Order.* Chicago: Chicago University Press.

Hayek F. A. 1952. *The Counter-Revolution of Science.* Indianapolis: Liberty.

Hayek F. A. 1967. *Studies in Philosophy, Politics and Economics.* London: Routledge and Kegan Paul.

Hayek F. A. 1978. *New Studies in Philosophy, Politics and Economics and the History of Ideas.* London: Routledge and Kegan Paul.

Hegel, G. W. F. 1977. *Phenomenology of Spirit.* Trans. A. V. Miller. Oxford: Clarendon Press.

Hegel, G. W. F. 1985. *Introduction to Lectures on the Philosophy of History.* Trans. T. M. Knox, and A.V. Miller, Oxford: Clarendon Press.

Heidegger, M. 1927. *Sein und Zeit.* Tübingen: Niemeyer.

Heidegger, M. 1962. *Being and Time.* Trans. Macquarrie, J. and Robinson, E. Oxford.

Heidegger, M. 2000. *Introduction to Metaphysics.* Trans. Fried, G., Polt, R. New Haven: Yale University Press.

Held, D. 1980. *An Introduction to Critical Theory.* Cambridge: Polity Press.

Hindess, B. 1977. *Philosophy and Methodology in the Social Sciences*. Sussex: Harvester Press.

Hirsch, E. D. 1976. *The Aims of Interpretation*. Chicago: Chicago University Press.

Hirsch, E. D. 1967. *Validity in Interpretation*. New Haven: Yale University Press.

Hollis, M. 1994. *The Philosophy of Social Science: An Introduction*. Cambridge: Cambridge University Press.

Hollis, M. 1996. *Reason in Action: Essays in the Philosophy of Social Science*. Cambridge: Cambridge University Press.

Hollis, M.; Lukes, S. eds. 1982. *Rationality and Relativism*. Oxford: Blackwell.

Hollis, N.; Nell, E. 1975. *Rational Economic Man*. Cambridge: Cambridge University Press.

Hookway, C.; Pettit, P. 1978. *Action and Interpretation: Studies in the Philosophy of the Social Sciences*. Cambridge: Cambridge University Press.

Horkheimer, M. 1968a. *Kritishhe Theorie: Eine Dokumentation*. Schmidt, A. ed. 2 vols. Frankfurt: S. Fischer Verlag.

Horkheimer, M. 1968b. 'Zum Rationalismusstreit in der gegenwärtigen Philosophie' [The dispute over rationalism in contemporary philosophy], *Kritische Theorie*, vol. 1.

Horkheimer, M. 1972. 'Art and Mass Culture', in *Critical Theory: Selected Essays*, trans. Matthew J. O'Connell et al., New York: Herder and Herder.

Horkheimer, M. 1974. *Critique of Instrumental Reason*. Trans. O'Connell, M. J. et al., New York Seabury Press.

Horkheimer and Adorno. 1979. *Dialectic of Enlightenment*. Trans. J. Cumming. London: Verso.

Hoy, D. (ed). 1986. *Foucault: A Critical Reader*. Oxford: Blackwell.

Hoy, D. 1985. 'Jacques Derrida' in *The Return of Grand Theory in the Human Sciences*, Skinner, Q., ed. Cambridge: Cambridge University Press, pp. 41–65.

Humboldt, W. von, 1971. *Linguistic Variability and Intellectual Development. Introduction to the Kawi Work*. Trans. Buck, G. C. and Raven, F. A. Coral Gables, FL: University of Miami Press.

Humboldt, W. von, 1968. *Gesammelte Schriften*. Leitzmann, A. et al. Prussian Academy of the Sciences. 17 Vols. Berlin: B. Behr, 1903–1916. Rpt. Berlin: de Gruyter.

Husserl, E. 1950–. Husserliana – *Gesammelte Werke*. The Hague.

Husserl, E. 1954. *Die Krisis der Europäischen Wissenschaften und die transzendentale Phänomenologie*. The Hague.

Husserl, E. 1970. *The Crisis of the European Sciences and Transcendental Phenomenology* trans. Carr, D. Evanston, IL: Northwestern University Press.

Husserl, E. 2001. *Logical Investigations*. Trans. Findlay, J. N., ed. Moran, D. London: Routledge.

Illich, I. 1993. *In the Vineyard of the Text: A Commentary on Hugh's Didascalion*. Chicago: Chicago University Press.

Ingarden, R. 1973. *The Literary Work of Art*. Evanston: Northwestern University Press.

Ingarden, R. 1975. *On the Motives Which Led Husserl To Transcendental Idealism*. Den Haag: Nijhoff.

Inwood, M. 1992. *Hegel Dictionary*. London: Blackwell.

Irenaeus, St.; Bishop of Lyons. 1920. *St. Irenaeus: The Demonstration of the Apostolic Preaching.* New York: Macmillan.

Jacoby, R. 1971. *'Towards a Critique of Automatic Marxism: The Politics of Philsoophy Form from Lukacs to the Frankfurt School'.* Telos, no. 10.

James, S. 1984. *The Content of Social Explanation.* Cambridge: Cambridge University Press.

Jameson, F. 1990. *Late Marxism: Adorno or the Persistence of the Dialectic.* London: Verso.

Jay, M. 1973. *The Dialectical Imagination: A History of the Frankfurt School and the Institute of Social Research, 1923–1950.* Boston: Little Brown.

Jay, M. 1984. *Adorno.* London: Fontana.

Jeanrond, Werner G. 1991. *Theological Hermeneutics.* London: Macmillan.

Joyce, P. 2002. *The Social in Question: New Bearings in History and the Social Sciences.* London: Routledge.

Kant, I. 1996. *Kant's First Critique and the Transcendental Deduction.* Aldershot: Avebury.

Kant, I. 1997. *Critique of Practical Reason.* Cambridge: Cambridge University Press.

Kant, I. 1977. 'Was Ist Aufklärung?' in Reiss, H. ed. *Kant's Political Writings* Cambridge: Cambridge University Press: 54–60.

Keat, R.; Urry, J. 1982. *Social Theory as Science.* London: Routledge and Kegan Paul.

Kockelmans, J. J. ed. 1967. *Phenomenology. The Philosophy of Edmund Husserl and its Interpretation.* Garden City, NY: Anchor Books Doubleday.

Kolakowski, L. 1978. *Main Currents of Marxism: The Founders, vol. 1.* Oxford: Oxford University Press.

Kortian, G. 1980. *Metacritique. The Philosophical Arguments of Jurgen Habermas.* Trans. J. Raffan. Cambridge: Cambridge University Press.

Kuhn, T. 1962. *The Structure of Scientific Revolutions.* Chicago: Chicago University Press.

Kukla, A. 2000. *Social Constructivism and the Philosophy of Social Science.* New York: Routledge.

Lakatos, I.; Musgrave, A. eds. 1970. *Criticism and the Growth of Knowledge.* Cambridge: Cambridge University press.

Lavrin, J. 1971. *Nietzsche: A Biographical Introduction.* London: Studio Vista.

Leibniz, G. W. 1991. *G. W. Leibniz's Monadology.* London: Routledge.

Lenin, V. I. 1972. 'The Three Sources and Component Parts of Marxism', in *Marx, Engels, Lenin: Selected Works.*

Lichtheim, G. 1961. *Marxism: An Historical and Critical Study.* London: Routledge and Kegan Paul.

Lukacs, G. 1971. *History and Class Consciousness.* Trans. R. Livingston. London: Merlin Press.

Luther, Martin. 1991. *Selected Readings from Martin Luther.* Alton: Hunt and Thorpe.

Mack, M. 2003. *German Idealism and the Jew: the Inner Anti-semitism of Philosophy and German Jewish Responses.* Chicago: University of Chicago Press.

Malthus, T. R. 1970. *An Essay on the Principle of Population. The First Essay.* Flew, A. ed. Harmondsworth: Penguin.

Manicas, P. T. 1987. *A History and Philosophy of the Social Sciences.* Oxford: Basil Blackwell.

Marcuse, H. 1976. 'On the Problem of the Dialectic'. Trans. Schoolman, M. and Smith, D. *Telos*, no. 27. Spring.

Martin, M.; McIntyre, L. C. 1994. *Readings in the Philosophy of Social Science*. Cambridge, MA: MIT Press.

Marx, K. 1964. *The German Ideology*. Moscow: Progress Publishers.

Marx, K. 1970. *Economic and Philosophic Manuscripts*. London: Lawrence and Wishart.

Marx, K. 1971. *A Contribution to the Critique of Political Economy*. Moscow: Progress Publishers; London: Lawrence and Wishart.

Marx, K. 1977a. *Economic and Philosophic Manuscripts*, in McLellan, D. ed. Selected writings (1977). Oxford: Oxford University Press.

Marx, K. 1977b. *The Poverty of Philosophy* in McLellan, D. ed. Selected writings (1977). Oxford: Oxford University Press.

Marx, K. 1992. *Communist Manifesto*. Oxford: Oxford University Press.

McNiece, G. 1992. *The Knowledge that Endures. Coleridge, German Philosophy and the Logic of Romantic Thought*. London: Macmillan.

Meister, J. G. 1698. *Dissertation on Interpretation*. Leipzig: Bay Mafin Theodor Heyben.

Mendelson, J. 1979. "The Habermas Gadamer Debate". *New German Critique* 18: 44–73.

Montaigne, Michel de. 1958. *The Complete Works*. London: Hamish Hamilton.

Montaigne, Michel de. 2004. *The Complete Essays*. London: Penguin.

Mueller-Vollmer, ed. 1985. *The Hermeneutics Reader: Texts of the German Tradition from the Enlightenment to the Present*. New York: Continuum.

Murray, M. ed. 1978. *Heidegger and Modern Philosophy: Critical Essays*. New Haven: Yale University Press.

Nehamas, A. 1985. *Nietzsche: Life as Literature*. Cambridge, MA: Harvard University Press.

Nietzsche, F. 1924. *Will to Power*. Trans. Ludovici, A. London: Allen and Unwin.

Nietzsche, F. 1961. *Thus Spoke Zarathustra*. Hollingdale, R. J. trans. Middlesex: Penguin Books.

Nietzsche. F. 1979. *Ecce Homo*, trans. Hollingdale, R. J. Harmondsworth: Penguin.

Nietzsche, F. 1990. *Beyond Good and Evil*. Trans. Hollingdale, R. J. Middlesex: Penguin Books.

Nietzsche, F. 1993. *The Birth of Tragedy*. Trans. Whiteside, S. Harmondsworth: Penguin Books.

Nietzsche, F. 1994. *Genealogy of Morals*. Ansell-Pearson, K. ed. Cambridge: Cambridge University Press (essay II).

Nietzsche, F. 1997. *Human, All to Human, in The Complete Works of Friedrich Nietzsche*, *Vol. 3, 1*. Stanford: Stanford University Press.

Nietzsche, F. 1997. *Untimely Meditations*. Trans. Hollingdale, R. J. Cambridge: Cambridge University Press.

O'Connor, B. 2004. *Adorno's Negative Dialectic: Philosophy and the Possibility of Critical Rationality*. Cambridge, MA: MIT Press.

Origen 1985. *On First Principles*. Trans. Butterworth. G. W. London: Peter Smith Publishers.

Outhwaite, W. 1987. *New Philosophies of Social Science: Realism, Hermeneutics and Critical Theory*. London: Macmillan.

Owensby, J. 1994. *Dilthey and the Narrative of History*. Ithaca, NY: Cornell University Press.

Palmer, R. 1969. *Hermeneutics: Interpretation Theory in Schleiermacher, Dilthey, Heidegger, Gadamer*. Evanston, IL: Northwestern University Press.

Parsons, T.; Loubser, J. J. 1976. *Explorations in General Theory in Social Science: Essays in Honour of Talcott Parsons, Vols. I & II*. New York: The Free Press.

Phillips, D. C. 1987. *Philosophy, Science, A Social Inquiry: Contemporary Methodological Controversy*. Oxford: Pergamon Press.

Pippin, R. 1989. *Hegel's Idealism: The Satisfaction of Self-Consciousness*. Cambridge: Cambridge University Press.

Porter, T. M.; Ross, D. 2003. *The Modern Social Sciences*. Cambridge: Cambridge University Press.

Potter, G. 2000. *The Philosophy of Social Science: New Perspectives*. Harvard: Pearson Education Ltd.

Pratt, V. 1978. *The Philosophy of the Social Sciences*. London: Methuen.

Quintillian, 2001. *The Orator's Education*. Cambridge, MA: Harvard University Press.

Rabinow, P. ed. 1991. *The Foucault Reader*. Harmondsworth: Penguin Books.

Ricoeur, P. 1976. *Interpretation Theory: Discourse and the Surplus of Meaning*. Fort Worth: Texas Christian University Press.

Ricoeur, P. 1981. *Hermeneutics and the Human Sciences: Essays on Language, Action and Interpretation*. Trans. Thompson, J. B. Cambridge: Cambridge University Press.

Ricoeur, P. 1984. *Time and Narrative*. Trans. McLaughlin, K. Pellauer, D. Chicago: University of Chicago Press.

Robinson, J. M. 1964. *The New Hermeneutic*. Robinson, J. M. and Cobb, J. B. ed. New York: Harper and Row.

Root, M. 1993. *Philosophy of Social Science: The Methods, Ideals, and Politics of Social Inquiry*. Oxford: Blackwell.

Rosen, M. 1982. *Hegel's Dialectic and Its Criticism*. Cambridge: Cambridge University Press.

Rosenberg, A. 1988. *Philosophy of Social Science*. Oxford: Clarendon Press.

Rosenberg, A. 2000. *Darwinism in Philosophy, Social Science, and Policy*. Cambridge: Cambridge University Press.

Rudner, R. S. 1966. *Philosophy of Social Science*. New York: Prentice-Hall Inc.

Runciman, W. G. 1969. *Social Science and Political Theory*. Cambridge: CUP.

Runciman, W. G. 1972. *A Critique of Max Weber's Philosophy of Social Science*. Cambridge: Cambridge University Press.

Ryan, A. 1970. *The Philosophy of the Social Sciences*. London: Macmillan.

Ryle, G. 1990. 'Knowing How and Knowing That' in *Collected Papers*, Bristol: Thoemmes Antiquarian Books. pp. 212–225.

Safranski, R. 2002. *Nietzsche: A Philosophical Biography*. London: Granta.

Saussure, F. de. 1916. *Cours de Linguistique Générale*, ed. Bally, C. and Sechehaye, A. Paris: Lausame.

Saussure, F. de. 1983. *Course in General Linguistics*. Trans. Harris R. London.

Sayer, D. 1979. *Marx's Method*. Brighton: Harvester.

Schleiermacher, F. D. E. 1959. *Hermeneutik*. Kimmerle, H. ed. Heidelberg: Carl Winter, Universitätsverlag.

Schleiermacher, F. D. E. 1977. *Hermeneutics: The Handwritten Manuscripts by F.D.E. Schleiermacher.* Trans. Kimmerle, H. ed., Duke, J. and Forstman J. Missoula, MT: Scholars Press.

Scott, G. 1991. *The History and Philosophy of Science.* London, New York: Routledge.

Sherratt, Y. 2002. *Adorno's Positive Dialectic.* Cambridge: Cambridge University Press.

Shoemaker, P. J.; Tankard, J. W.; Lasorsa, D. 2004. *How to Build Social Science Theories.* London: Sage.

Skinner, Q. 1988. *Meaning and Context: Quentin Skinner and his Critics.* Cambridge: Polity Press.

Soll, I. 1969. *An Introduction to Hegel's Metaphysics.* Chicago: Oxford University.

Solomon, R. 1983. *In the Spirit of Hegel.* Oxford: Oxford University Press.

Staiger, E. 1991. *Basic Concepts of Poetics.* Trans. Hudson, J. Frank, L. Philadelphia: Pennsylvania State University Press.

Stretton, H. 1969. *The Political Sciences: General Principles of Selection in Social Sciences and History.* London: Routledge and Kegan Paul.

Taylor, C. 1985. 'Foucault on Freedom and Truth' in *Philosophy and the Human Sciences: Philosophical Papers Volume 2.* Cambridge: Cambridge University Press.

Thomas, D. 1979. *Naturalism and Social Science: A Post-Empiricist Philosophy of Social Science.* Cambridge: Cambridge University Press.

Thompson, J. B. 1981. *Cultural Hermeneutics: A Study in the Thought of Paul Ricoeur and Jurgen Habermas.* Cambridge: Cambridge University Press.

Torrance, Thomas F. 1988. *The Hermeneutics of John Calvin.* Edinburgh: Scottish Academic Press.

Trigg, R. 2001. *Understanding Social Science.* Oxford: Basil Blackwell.

Tudor, A. 1982. *Beyond Empiricism: Philosophy of Science in Sociology.* London: Routledge and Kegan Paul.

Turner, S. P. 1986. *The Search for a Methodology of Social Science: Durkheim, Weber, and the Nineteenth Century Problem of Cause, Probability and Action.* Dordrecht: D. Reidel.

Uffenhimer, B. and Reventlow, H. G. 1988. *Creative Biblical Exegesis: Christian and Jewish Hermeneutics through the Centuries.* Tel Aviv: JSOT Press.

Vattimo, G. 2002. *Nietzsche: An Introduction.* London: Athlone Press.

Vico. 1982. *Selected Writings,* ed. and trans. Pompa, L. Cambridge and London: Cambridge University Press.

Wagner, P. 2001. *A History and Theory of the Social Sciences: Not All That Is Solid Melts into Air.* London: Sage.

Walsh, W. H. 1985. 'The Origins of Hegelianism' in Inwood, M. *Hegel.* Oxford: Oxford University Press, pp. 13–30.

Warnke, G. 1987. *Gadamer: Hermeneutics, Tradition and Reason.* Cambridge: Polity Press.

White, S. K. ed. 1995. *The Cambridge Companion to Habermas.* Cambridge: Cambridge University Press.

Whitebook, J. 1995. *Perversion and Utopia: A Study in Psychoanalysis and Critical Theory.* Cambridge, MA: MIT Press.

Wiggershauss, R. 1994. *The Frankfurt School: Its History, Theories and Political Significance.* Trans. Robertson, M. Cambridge: Polity Press.

Winch, P. 1958. *The Idea of Social Science.* London: Routledge and Kegan Paul.

Wisdom, J. O. 1993. *Philosophy of the Social Sciences.* Aldershot: Avebury.

Wohlfart, G. 1981. *Der Spekulative Satz: Bermerkungen zum Begriff der Speulation bei Hegel.* Berlin: de Gruyter.

Wolff, C. 2003. *Logic, or Rational Thoughts on the Powers of the Human Understanding.* Trans. Wolfias, B. New York: Olms.

Worral, J. 1996. 'Is the Idea of Scientific Explanation Unduly Anthropocentric?: The Lessons of the Anthropic Principle.' *LSE Centre for the Philosophy of the Natural and Social Sciences Discussion Paper Series.* London: Tymes Court.

Wren, T. E.; Edelstein, W. 1990. *The Moral Domain: Essays in the Ongoing Discussion between Philosophy and the Social Sciences.* Cambridge, MA: MIT Press.

Zuidervaart, L. 2004. *Artistic Truth.* Cambridge: Cambridge University Press.

Index